BRAIN-EATING AMOEBAE

Biology and Pathogenesis of *Naegleria fowleri*

Ruqaiyyah Siddiqui[1]
Ibne Karim M. Ali[2]
Jennifer R. Cope[2]
Naveed Ahmed Khan[1]

[1]Sunway University, Selangor, Malaysia
[2]Centers for Disease Control and Prevention, Atlanta, USA

ISBN: 978-1-910190-53-1 (paperback)
ISBN: 978-1-910190-54-8 (ebook)

Dedication

To our beloved children, Salahuddin Ahmed Khan, Mohammad Hafeez Khan and Imaan Asadi Khan, who bring so much joy to our lives.

Ruqaiyyah Siddiqui and Naveed Ahmed Khan

Contents

Preface

The purpose of this book is to provide a reference for *Naegleria fowleri* as a quick guide for clinical colleagues, health professionals, researchers, and students. It is purposefully kept brief, and is divided into easy to follow sections, covering all aspects of *N. fowleri*. This compilation will serve as an essential reference for microbiologists, immunologists, physicians, and public health officials, in the field of basic and medical microbiology, as well as an invaluable reference for new and experienced researchers who wish to understand this organism better. This book is the definitive guide to the current knowledge and ongoing research in this medically and ecologically important organism. We are indebted to Drs. Jennifer R. Cope and Ibne Karim M. Ali (Waterborne Disease Prevention Branch in the National Center for Emerging and Zoonotic Infectious Diseases), for superbly composing the chapters on current practices and advances in diagnostics and therapeutics against *N. fowleri* infection.

We are most grateful to Professor Govinda Visvesvara (Centers for Disease Control, Atlanta, GA, USA), and Professor Rolf Michel (Department of Microbiology / Parasitology, Central Institute of the Federal Armed Forces Medical Service, Koblenz, Germany) and Dr. Bret S. Robinson, Australian Water Quality Centre, South Australian Water Corporation) for kindly providing the images used in this book and their encouraging comments.

We are very thankful to colleagues who read, commented upon chapters or sections, for accuracy, including Dr. Sutherland Maciver (University of Edinburgh), Professor Govinda Visvesvara (Centers for Disease Control, Atlanta, GA, USA), and Professor Ed Jarroll (City University New York). Our special thanks to Dr. Sutherland Maciver for his never ending support, helpful discussions, and spending so much time with dedication in completing this project. Finally, we are thankful to Sunway University for the freedom and understanding to embark on this project.

Ruqaiyyah Siddiqui and Naveed Ahmed Khan
Sunway University

Synopsis

Naegleria fowleri is a eukaryotic protist pathogen that causes primary amoebic meningoencephalitis. It enters the brain via the nasal route and kills the host within days. The most distressing aspect is that the disease almost always results in death with more than 90% mortality rate. Despite our advances in antimicrobial chemotherapy and supportive care, it is considered as one of the world's deadliest known parasites. This is, in part due to our incomplete understanding of the biology, ecology, pathogenesis, pathophysiology, and lack of available effective drugs. A complete knowledge of this parasite, how it lives in the environment, and produces disease is crucial for the rational development of preventative and therapeutic strategies against this fatal, albeit rare disease.

Current books of interest

Introduction to *Naegleria*

The genus *Naegleria* is a group of protist organisms that are widely distributed in the environment. Among the different species, *N. fowleri* is the only species that is known to infect people by entering the body through the nose via contaminated water, and produces primary amoebic meningoencephalitis, which is almost always fatal. *N. fowleri* has a three stage life cycle, consisting of a trophozoite stage, a flagellate form, and a cyst stage (Figure 1). It is widely accepted that the trophozoite stage is the only infective stage of the amoeba. *N. fowleri* reproduces asexually by binary fission during the trophozoite stage. In the absence of food but presence of water, *N. fowleri* trophozoites transform into the flagellate form to travel long distance in search of nutrients. The flagellate form is a transient stage that can neither encyst, nor reproduce. Under harsh conditions, trophozoites transform into the cyst form. The cyst is a dormant stage during which they remain inactive with little metabolic activity, but remain viable, for years. Both the flagellate and the cyst are non-feeding, and non-reproductive forms. Only the trophozoite form is able to feed, divide, and encyst. Cysts will excyst to yield trophozoites under favourable environmental conditions. When studying biology and pathogenesis, *N. fowleri* has often been compared against the well-studied non-pathogenic *N. gruberi*, however it is considered more appropriate to compare *N. fowleri* against closely related non-pathogenic *N. lovaniensis*. Given the opportunity and access, *N. fowleri* infect humans via the nose during swimming, nasal cleansing, bathing etc. and enter the brain via the olfactory neuroepithelial route to produce brain infection. The true burden of primary amoebic meningoencephalitis due to *N. fowleri* on human health is not known, as the majority of infections in less developed countries go unnoticed and in many developing countries the public has limited access to clean water. Furthermore, the pathogenesis and pathophysiology associated with *N. fowleri* infection, as well as the molecular identification of virulence traits of *N. fowleri*, which will be potential targets for therapeutic interventions, and/or the development of preventative strategies remain incompletely understood.

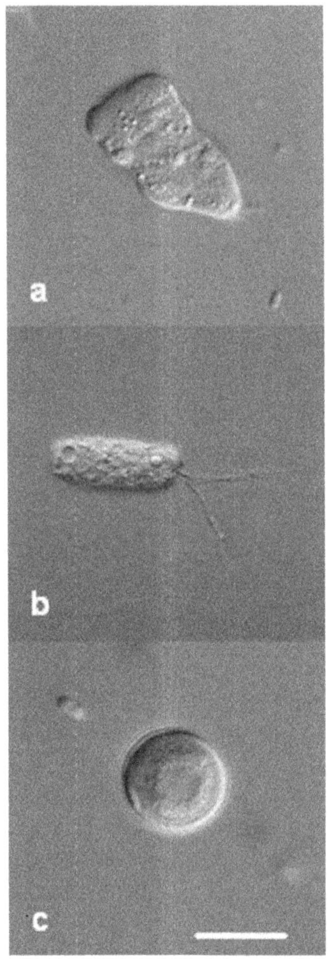

Figure 1. *N. fowleri* (a) trophozoite form, (b) flagellate form, and (c) cyst form (courtesy: B. S. Robinson, Australian Water Quality Centre, South Australian Water Corporation). Bar is 10µm.

1

Primary Amoebic Meningoencephalitis

Abstract

Since the discovery of *N. fowleri* several decades ago, major efforts have failed to prevent and successfully treat primary amoebic meningoencephalitis. It is considered as one of the most aggressive parasitic infections, that almost always leads to death within a few days. The clinical course is dramatic exhibiting headache, stiff neck, seizures, coma, and death. The trophozoite enters the nose via contaminated water or dust, travels along the olfactory neuroepithelial route to reach the central nervous system and to provoke haemorrhagic necrosis. With the devastating nature of this disease and problems associated with its chemotherapy, here we describe the clinical features, pathophysiology and risk factors associated with primary amoebic meningoencephalitis.

1.1. The disease

N. fowleri produces an acute fulminant, haemorrhagic-necrotizing meningoencephalitis, charact-erised by severe headache, stiff neck, fever (38.5°C - 41°C), altered mental status, seizures, and coma, leading almost always to death. Primary amoebic meningoencephalitis is associated with an inflammatory reaction composed of neutrophils, eosinophils, macrophages and a few lympho-cytes. The incubation period from exposure leading to meningoencephalitis may range from one to 16 days, depending partly on the size of the inoculum and virulence of the strain (Martinez, 1977). Primary amoebic meningo-encephalitis follows inhalation of water containing amoebae or flagellates. Parasites penetrate the nasal mucosa and the cribiform plate and travel along the olfactory nerves (Symmers, 1969). The olfactory nerve terminates in the olfactory bulb, which is located in the richly vascularized subarachnoid space and is bathed by cerebrospinal fluid (Martinez, 1977). The subarachnoid space is the route of dissemination to the rest of the central nervous system. Although a retrospective study suggested that primary amoebic meningoencephalitis infections most probably caused by *N. fowleri* had occurred in Virginia in 1937 (Dos Santos, 1970), the first definite reported infection of primary amoebic meningoencephalitis was described in 1965 in Australia (Fowler and Carter, 1965). To date, a few hundred cases of primary amoebic meningoencephalitis have been reported worldwide, with most cases reported in the United States, Australia, and Europe (France). Trophozoites are the only form found in the lesions (Figure 1.1). Amoebae first invade the olfactory bulbs and then spread to the more posterior regions of the brain. *N. fowleri* has been isolated from the cerebrospinal fluid of infected patients (Cerva et al., 1969; Warhurst et al., 1970). Within the brain they provoke inflammation (Rojas-Hernández et al., 2004b) and cause extensive damage to the tissue. The clinical course is dramatic. Symptoms begin with severe frontal headache and fever (38.5 - 41°C). This is followed by nausea, vomiting, and

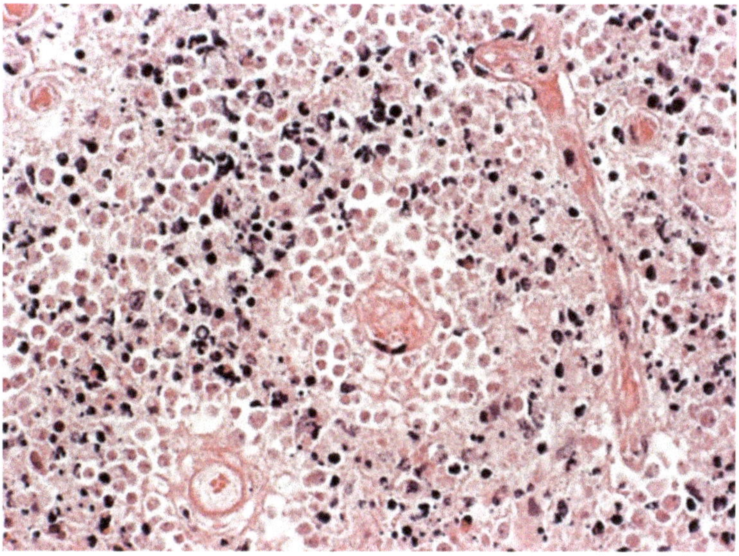

Figure 1.1. A section of the cerebral portion of the brain of *N. fowleri*-infected patient, stained with hematoxylin and eosin, showing *N. fowleri* trophozoites, X400, (courtesy: G. S. Visvesvara, CDC, USA).

signs of meningeal irritation. Involvement of the olfactory lobes may cause: disturbances in the sense of smell or taste and may be noted early in the course of the disease. Visual disturbances may occur. The patient may experience confusion, irritability, and restlessness and may become irrational before lapsing into coma. Generalized seizures also may be present. In order of frequency of occurrence, the more important symptoms include headache, pyrexia, nausea, vomiting, fever, and stiff neck (John, 1982). The disease is designated "primary" to distinguish it from meningocerebral infection caused by parasitic amoebae, *Entamoeba histolytica*, which invade the central nervous system only as a result of dissemination in the bloodstream from lesions in other parts of the body.

1.2. Pathology

N. fowleri invades humans via nasal mucosa, crosses the cribriform plate, migrates along the basilar brain from the olfactory bulbs and tracts to the cerebellum, deeply penetrates the cortex to the periventricular system, and incites meningoencephalitis with rapid cerebral oedema, resulting in cerebellar herniation. The olfactory bulbs and orbitofrontal cortices are necrotic and haemorrhagic (Figure 1.2). The cerebral hemispheres usually are oedematous. Meninges are diffusely hyperemic with a slight purulent exudate. The cortex contains many focal superficial haemorrhages. The olfactory bulbs exhibit marked involvement with haemorrhage, necrosis, and purulent exudate. The leptomeninges (arachnoid and pia mater) are severely congested, diffusely hyperemic,and opaque with limited purulent exudate within sulci, base of the brain, brainstem, and cerebellum. Trophozoites, but not cysts, are usually seen within the Virchow-Robin spaces. Amoebae, ranging in size from 8 to 12 µm and possessing densely staining round nuclei, are found in large numbers in the base of the brain, hypothalamus, and midbrain. The scans of the brain show obliteration of the cisternae around the midbrain and the subarachnoid space over the cerebral hemispheres.

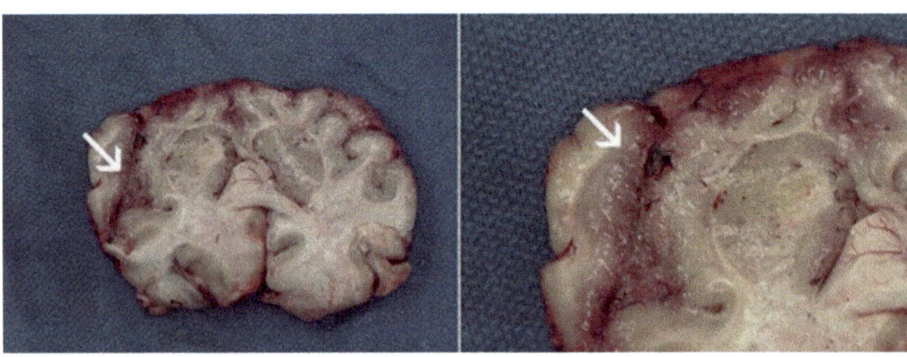

Figure 1.2. *N. fowleri*-infected brain exhibiting extensive haemorrhage and necrosis in the frontal cortex (courtesy: G. S. Visvesvara, CDC, USA).

Marked diffuse enhancement in these regions may be seen after administration of intravenous contrast medium. Histology has shown acute inflammatory reaction, mainly composed by neutrophils with extensive areas of lytic necrosis with the presence of several trophozoites. The literature in human and in experimental mouse models showed that the fibrinopurulent exudate is practically absent. Increased intracranial pressure and herniation are usually the cause of death. In the advanced stage, the red blood cells increase up to 24,600 per mm^3. The white blood cell count (predominantly polymorphonuclear leukocytes), varies from 300 cells per mm^3 to as high as 26,000 per mm^3. The protein concentration ranges from 100 mg to 1000 mg per 100 mL, and glucose may be 10 mg per 100 mL or lower (reviewed in Visvesvara, 2013).

Microscopic examination reveals many amoebae in the subarachnoid and perivascular spaces (Figure 1.3). Presumably, the perivascular spaces provide a path of migration for the amoebae, and the blood vessels supply the oxygen needed by these aerobic organisms. In fewer numbers, amoebae are found clustered within the brain tissue and in the purulent exudate of the meninges and brain substance (reviewed in John, 1982). Within the exudate some amoebae may be seen engulfed by macrophages. Many amoebae are observed to contain phagocytosed cellular debris and erythrocytes. The purulent exudate contains numerous polymorphonuclear and mononuclear leukocytes. The cortical gray matter is a preferred site for amoebae development; consequently, severe involvement occurs in the cerebral hemispheres, cerebellum, brain stem, and upper portions of the spinal cord. Encephalitis ranges from slight amoebic invasion and inflammation to massive invasion with purulent, haemorrhagic necrosis. Typically, the olfactory bulbs exhibit extensive amoebic invasion, haemorrhage, and an inflammatory exudate; the involvement here is greater than in other areas of the brain. Focal

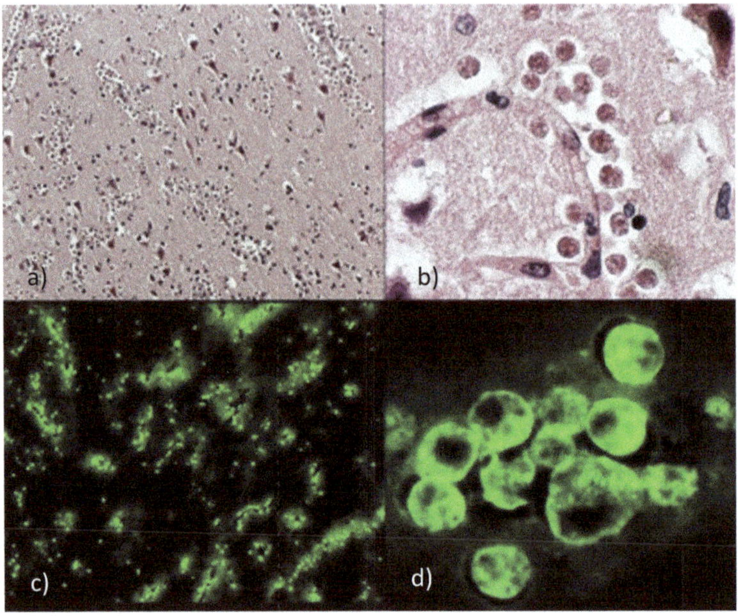

Figure 1.3. A section of the cerebral portion of the brain of *N. fowleri*-infected patient, stained with hematoxylin and eosin (a and b), and immunofluorescent staining using anti-*N. fowlerl* antibody (c and d), showing clusters of *N. fowleri* trophozoites (cysts are not seen) (courtesy: G. S. Visvesvara, CDC, USA).

demyelination in the white matter of the brain and spinal cord has been described. Demyelination occurred in the absence of amoebae or cellular infiltration. Chang (1979) suggests that demyelination is caused by a phospholipolytic enzyme or enzyme-like substance produced by actively growing amoebae present in the adjacent gray matter. Foci of neuronal degeneration and demyelination have been described (Viriyavejakul et al., 1997). Myocarditis has been described in some patients dying of primary amoebic meningoencephalitis (Markowitz et al., 1974).

1.3. Risk factors

To date, a few hundred cases of primary amoebic meningoencephalitis have been reported worldwide, with most cases reported in the United States, Australia, and Europe (France). As the organism lives and multiplies in warm water, including freshwater lakes, river, canals, geothermal springs, spas, untreated domestic water supplies and swimming pools, a major risk factor is swimming in the contaminated water. This is consistent with the fact that typically, primary ameobic meningoencephalitis occurs in healthy, young individuals, who frequently have a history of swimming or washing their face in infested waters. In addition, prolonged hot and dry periods due to global warming are causing higher freshwater temperatures that are coinciding with augmented amoebal densities in water supplies, as well as an increase in recreational activities that are likely attributing to a rise in primary amoebic meningoencephalitis. Although primary ameobic meningoencephalitis might occur more frequently in tropical regions, the majority of known cases are reported from subtropical or even temperate zones. The fact that the infection is underreported in tropical regions is probably due to better public health services in the temperate zones, while in tropical zones primary amoebic meningoencephalitis infections might go unnoticed among the millions of other infections. For example, in countries such as Pakistan, temperatures can reach up to 50°C, while water temperatures are recorded at 30 - 35°C, and with prolonged power cuts, millions of people turn to freshwater canals, ponds, standing water etc. Large crowds take part in this "recreational activity" which goes on almost every day for months, during the summer period (Figure 1.4). Lack of available toilet facilities and use of same waters to defecate with no apparent signage for potential dangers of swimming is both noticeable and disturbing. The presence of *N. fowleri* in these waters, lack of awareness and/or control measures, poor healthcare infrastructure, and

Figure 1.4. Thousands of people can be seen swimming in the canal that passes through the city of Lahore, Pakistan, without nearby facilities for defecation and urination (from PLoS Negl Trop Dis. 2014, e3017).

unavailability of effective drugs to counter this infection presents a major health hazard for the community.

In recent years, primary amoebic meningoencephalitis has been linked with ablution practice. Muslims pray five times a day. Before every prayer, they perform ablution for cleansing. This involves washing the hands, mouth, nose, ears, face, arms and feet. When cleaning the nose, many people push water forcefully up the nostrils, even though this is not a mandatory part of the ablution process. Although the ablution practice has tremendous health benefits, but it can only serve its purpose if water supplies are free of pathogenic microbes. In the presence of *N. fowleri* and if water is pushed up the nose, the deadly amoebae access the nasal mucosa to invade the brain and cause this lethal infection (Siddiqui and

Khan, 2014) (Figure 1.5). This suggests that rigorous ablution is an important risk factor in contracting primary amoebic meningoencephalitis. Thus there is a need for awareness to take measures to make water safer for ritual nasal rinsing. Using sterile water that is boiled for at least one minute and left to cool, or water filtered to remove small organisms, or water disinfected appropriately using recommended concentrations of chlorine together with careful ablution (not pushing water inside nostrils vehemently) should minimize the risk in contracting primary amoebic meningoencephalitis infection. In many developing countries, water scarcity is a major problem and public has to store water in tanks for days to weeks for their routine consumption which presents a major risk factor. The general

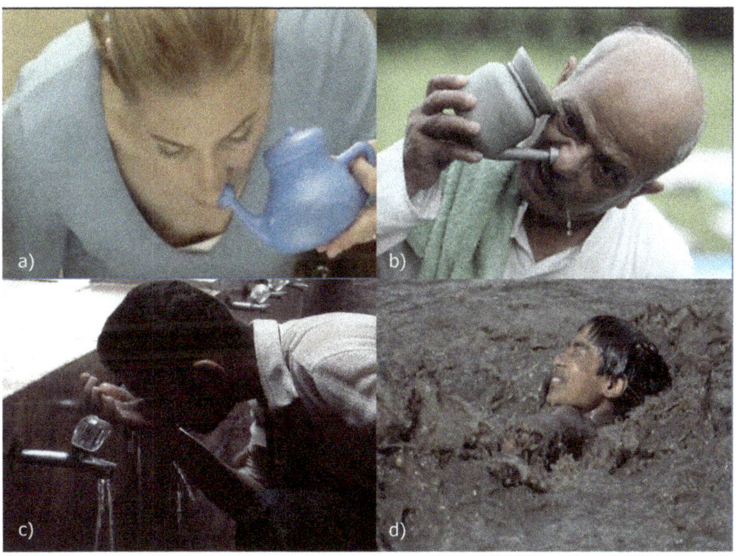

Figure 1.5. Nasal cleansing/irrigation using neti pots can provide relief to patients with sinusitis by flushing out excess mucus and debris from the nose (a and b). Ablution involves nasal cleansing (c). The use of contaminated water together with forceful pushing up the nostrils even though it is not required as part of the ablution practice or swimming in contaminated water (d) or unchlorinated pools with amoebae can lead to parasite entry into the brain.

public needs to be made aware of the health risks associated with the use of storage tanks at home and at prayer places. More importantly, the public needs to be informed of the appropriate maintenance of water storage tanks together with support, as occurrence of amoebae in domestic tap water supplies is an important public health risk. Apart from homes, inappropriate maintenance of water storage in mosques can also contribute to microbial contamination. As the majority of public perform ablution at mosques in preparation for prayers, it is imperative that water storage tanks are cleaned and disinfected (reviewed in Siddiqui and Khan, 2014). Other religious festivals such as the Kumbh Mela, where millions of Hindus gather in the Indian city of Allahabad for a ritual bath in the sacred Ganges River pose a risk to public health in the transmission of infectious agents (Siddiqui and Khan, 2014). In this month-long festival, the bathing takes place in an area known as the Sangam at the confluence of the Ganges and Yamuna rivers and a third mythical waterway called the Saraswati, and up to 100 million people participate in this holy bathing festival. At present, there is neither a report of primary amoebic meningoencephalitis-associated with this practice nor the prevalence of *N. fowleri* in these waters, and this should be investigated in future studies.

Nasal cleansing/irrigation using neti pots are often used to provide relief to patients with sinusitis including symptoms of facial pain, headache, cough, rhinorrhea (allergic rhinitis) and nasal congestion. Routine nasal cleansing can reduce medication used by patients with sinusitis and provide relief for hay-fever, common cold, and other chronic sinus and nasal symptoms. In addition, nasal irrigation is used in the practices such as Ayurveda, also known as "jala neti,", which involves sniffing water from cupped hands and then blowing it out (Figure 1.5). Nasal irrigation is performed using a device shaped like Aladdin's lamp-shaped device that is filled with saline (reviewed in Siddiqui and Khan, 2014). The water flows out of the tip of the pot into one nostril. Gravity takes the water around the back of the nostril and actually drains out the opposite side of the nose. Then the same procedure is repeated on the opposite side. Although nasal irrigation promotes good sinus and nasal health, it can only be effective if purified, filtered, or boiled water is used. The use of unboiled or otherwise unsterilized water has proven to be an important risk factor in contracting primary amoebic meningoencephalitis (reviewed in Siddiqui and Khan, 2014). Given the widespread use of this practice globally and the lack of availability of clean (sterilized/filtered) water to the majority of the population in the developing countries, it is likely that a large number of primary amoebic meningoencephalitis cases go unnoticed. This highlights the importance of raising awareness about this disease among physicians as well as the community.

The majority of primary amoebic meningoencephalitis cases have been reported in young males with a history of exposure to contaminated water. Considering recreational activities, this could be logically explained by fervent involvement of young men in outdoor activities. With regards to ablution, all primary amoebic meningoencephalitis cases at Aga Khan University Hospital have been observed in Muslim young males, and not a single case has been reported in women. Although the virulent nature of *N. fowleri* combined with rigorous ablution practices by males may be a contributing factor, it is likely that other predisposing factors may play a role in contracting primary amoebic meningoencephalitis infection. Thus there is need to reveal causes, be that (i) genetic, (ii) biochemical changes, or (iii) underlying disease prerequisite leading to damaged or abnormal mucosa (containing abnormal levels of immune factors), and/or (iv) due to rigorous, routine nasal irrigation. Future research is needed to address these issues. The use of nose clips should be encouraged to avoid any traumatic disruptions in the nasal mucosal linings during water-related activities in warm freshwater, such as lakes, rivers, ponds, bayous, and hot springs. When performing ablution, water that is boiled for one minute and left to cool, or filtered to remove small organisms, or disinfected appropriately should be utilized, together with cautious ablution (not pushing water inside nostrils vehemently), in order to avoid serious consequences for communities living in developing countries. Given the rapid onset and progression of primary amoebic meningoencephalitis in humans and the route of infection, future studies are needed to determine precise host factors to help develop preventative strategies as well as therapeutic interventions.

2

Clinical and Laboratory Diagnosis

Abstract

As the clinical course of primary amoebic meningoencephalitis is rapid, prompt detection of amoebae is crucial to increase the likelihood of patient survival by initiating early treatment. In general, due to non-specific and overlapping signs and symptoms of primary amoebic meningoencephalitis with those of bacterial or viral meningitis, and the rarity of *N. fowleri* infections, most primary amoebic meningoencephalitis cases are diagnosed retrospectively, i.e., during the autopsy examination of brain tissues. However, in live patients, primary amoebic meningoencephalitis is commonly diagnosed through microscopic examination of cerebrospinal fluid specimens. It is time critical to diagnose primary amoebic meningoencephalitis and begin empiric antimicrobial therapy. The diagnosis of primary amoebic meningoencephalitis depends on clinical features together with microscopic and/or molecular identification of the parasites in the cerebrospinal fluid specimens. This overview presents the diagnostic approach to primary amoebic meningoencephalitis.

2.1. Computed tomographic (CT) appearance

The computed tomographic (CT) appearance of primary amoebic meningoencephalitis due to *N. fowleri* reveal the cisterns around and above the midbrain and the subarachnoid spaces are obliterated on pre-contrast CT. Marked enhancement in these regions are seen after intravenous contrast medium administration. The sulci and adjacent gray matter are also strongly enhanced. The ventricular size is normal. Pathological findings are those of arachnoiditis and invasion of the leptomeninges and brain substance by amoebae, especially at the base of the brain and cerebellum (Lam et al., 1982).

2.2. Clinical specimens

The advantages and disadvantages of some of the commonly used diagnostic methods are provided in Table 2.1. In addition to a history of swimming or use of contaminated water for nasal cleansing, nasal discharge should be collected from patients exhibiting symptoms of primary amoebic meningoencephalitis. Fresh microscopic observation of the exudates in 0.85% sodium chloride could reveal active amoeba together with flagellate forms (Siripanth et al., 2005). Additionally, immunofluorescence assays and PCR-based assays should be employed for sensitive and specific detection. The best specimens from acute primary amoebic meningoencephalitis patients are cerebrospinal fluid and brain tissue (especially surrounding the olfactory bulb of the brain). These specimens should be collected aseptically, and kept at room temperature (~25°C) prior to any laboratory examinations. These conditions will allow the thermophilic amoebae to survive in the specimens, facilitating direct observation of live amoebae upon microscopic examination (Figure 2.1), and enabling growth in appropriate culture media. Although no case of nosocomial infection has been reported yet, personnel handling these specimens should wear appropriate protective equipment such as gloves and surgical masks, and should open the containers containing

Table 2.1. Advantages and disadvantages of different methods used in diagnosis of primary amoebic meningoencephalitis due to *N. fowleri*.

Methods	Advantages	Disadvantages
Microscopy	Simple, rapid, inexpensive	Familiarity and expertise in the identification of amoebae is required; detection is dependent on the morphological integrity of amoebae on the test specimens; non-specific as morphologically similar other *Naegleria* species exist
Culture	Aids in the confirmation of infecting species; helps n downstream research work requiring large number of amoebae, such as next generation sequencing or proteomic analyses	Not always successful as amoebae may not grow in the culture; time consuming; laborious; needs expertise to maintain the amoeba cultures; expensive
Serology	Serologic response against *N. fowleri* has been detected in only a handful of primary amoebic meningoencephalitis survivors	Not very helpful for diagnostic purposes as the disease progression is too rapid to mount a serologic response; cannot differentiate between an old exposure and an acute infection
Antigen detection	Species-specific when *N. fowleri* specific mAbs are used; does not require expertise on the amoeba morphology; relatively quick; relatively simple to perform; may work in variety of sample types	Not fully tested on clinical specimens except for an immunofluorescence based antigen detection test used at CDC on clinical specimens; sensitivity is lower than the PCR based tests
PCR (conventional, nested and real-time PCR)	Rapid, specific, highly sensitive; works on CSF as well as tissue biopsy or autopsy samples; requires small volume of sample; may provide genotype information	Requires sophisticated instruments and expertise; sometimes prone to contamination (especially the conventional PCRs)

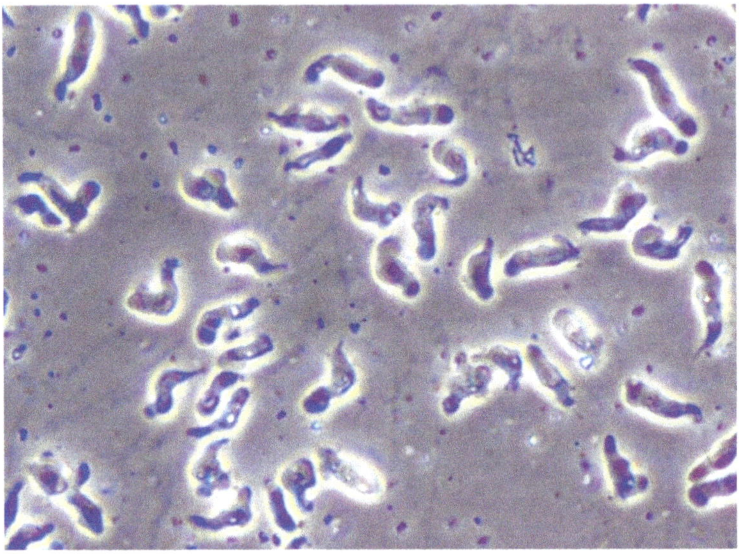

Figure 2.1. A wet mount of *N. fowleri* trophozoites cultured from the cerebrospinal fluid of a patient with primary amoebic meningoencephalitis viewed using phase contrast microscopy. X600. (courtesy: G. S. Visvesvara, CDC, USA).

the specimens only inside a biological safety cabinet to minimize risk of unintentional infection.

2.3. Microscopic identification of amoebae

2.3.1. Cerebrospinal fluid

Cerebrospinal fluid is presently the most useful and widely used specimen for primary amoebic meningoencephalitis diagnosis (Table 2.2). The cerebrospinal fluid profile in primary amoebic meningoencephalitis closely resembles that seen in bacterial meningitis. The cerebrospinal fluid colour from primary amoebic meningoencephalitis patients varies from one patient to another, and ranges from greyish to yellowish-white, or can be tinged red. As disease progresses, the red blood cell count in cerebrospinal fluid increases several fold from 250 cells per mm^3 in the early stage to

Table 2.2. Characteristics of cerebrospinal fluid (CSF) in primary amoebic meningoencephalitis patients.

Features	**Characteristics**	**Possibility of differential diagnosis?**
Colour	Greyish to yellowish-white, or tinged red	No
Number of RBCs	250 cells/mm^3 in early stage to 25000 cells/mm^3 in late stage	No
WBC count	PMN: 300 cells/mm^3 to 26000 cells/mm^3	No
Pressure	300-600 mmH_2O	Maybe; elevated in primary amoebic meningoencephalitis patients
Protein concentration	100 mg/100 mL to 1000 mg/100 mL	Maybe; elevated in primary amoebic meningoencephalitis patients
Glucose concentration	10 mg/100 mL or less	No
Presence of amoebae	Trophozoites of *N. fowleri* are often seen in acute cases	Yes

25,000 cells per mm^3 in the late stage. Similarly, the white blood cell count is elevated, with a polymorphonuclear leukocyte predominance, with a range of 300 cells per mm^3 to as high as 26,000 cells per mm^3. The cerebrospinal fluid pressure is usually elevated (300 - 600 mm H_2O). The protein concentration can range from 100 mg per 100 mL to 1000 mg per 100 mL, and glucose may be 10 mg per 100 mL or lower (Visvesvara, 1985; Visvesvara and Moura, 2006). In the majority of cases, motile *N. fowleri* trophozoites are seen in cerebrospinal fluid by wet mount. A wet-mount of the cerebrospinal fluid should be examined immediately after collection under a microscope (preferably equipped with phase-contrast optics) for the presence of actively moving trophozoites. If an immediate microscopic examination is not possible, then the cerebrospinal fluid may be stored at room temperature (~25°C) until it is viewed under the microscope or shipped to a reference laboratory. A gentle agitation of the container holding the cerebrospinal fluid may be helpful to dislodge any amoebae that are adhered to the container before observation under the microscope. Also, brief centrifugation of the cerebrospinal fluid specimen at 5,000 x g for 5 min may be helpful to concentrate amoebae at the bottom of the container prior to placing a small volume on a microscopic slide. Following centrifugation, the supernatant should be carefully removed, leaving about 200 - 300 µL of residual liquid without dislodging any visible pellet. The microscopic slide containing the cerebrospinal fluid may be placed inside a 35 - 37°C incubator to warm it, which facilitates the movement of any *N. fowleri* amoebae. Smears of cerebrospinal fluid should be stained with Giemsa or Wright stains to identify the trophozoite of *N. fowleri*. The amoeba can be clearly differentiated from host cells by the nucleus with its centrally placed large nucleolus. A Gram stain is generally not useful.

2.3.2. Brain tissue

Brain tissue from a suspected primary amoebic meningo-encephalitis patient obtained via biopsy or post-mortem exam should be examined microscopically. Fibrino-purulent leptomeningeal exudate will contain predominantly polymorphonuclear leukocytes, eosinophils, macrophages and some lymphocytes in the cerebral hemispheres, brain stem, cerebellum, and upper portion of the spinal cord. In most cases, a large number of *N. fowleri* trophozoites are seen usually in different pockets within the oedematous and necrotic neural tissue. Trophozoites of *N. fowleri* are also seen in the Virchow-Robin spaces, usually around the blood vessels with no inflammatory response. Trophozoites of *N. fowleri* range in size from 8 to 12 µm, and are recognizable by the presence of a large nucleus with a centrally located large nucleolus. No *N. fowleri* cysts are seen in the brain tissues. *N fowleri* trophozoites proliferate in brain tissue, meninges, and the cerebrospinal fluid.

2.3.3. Amoebic culture

N. fowleri can be cultured from cerebrospinal fluid or brain tissue samples. One or two drops of cerebrospinal fluid or a small portion of brain tissue may be placed onto an agar plate overlaid with bacteria or directly into axenic growth medium. They may also be inoculated onto tissue culture monolayers for amoebic growth. Growth of amoebae is checked every day for seven days. Usually, trophozoites are easily seen in the plate or growth media within 48 h (Figure 2.2). Amoebic culture takes several days to complete. As a result, it has a very low to no diagnostic value but may be used to generate large number of amoebae as needed for research purposes. For isolation and culture of *N. fowleri* in non-nutrient agar, it is important to note that neither physiological saline nor phosphate buffered saline are useful as the concentration of sodium chloride in these solutions is

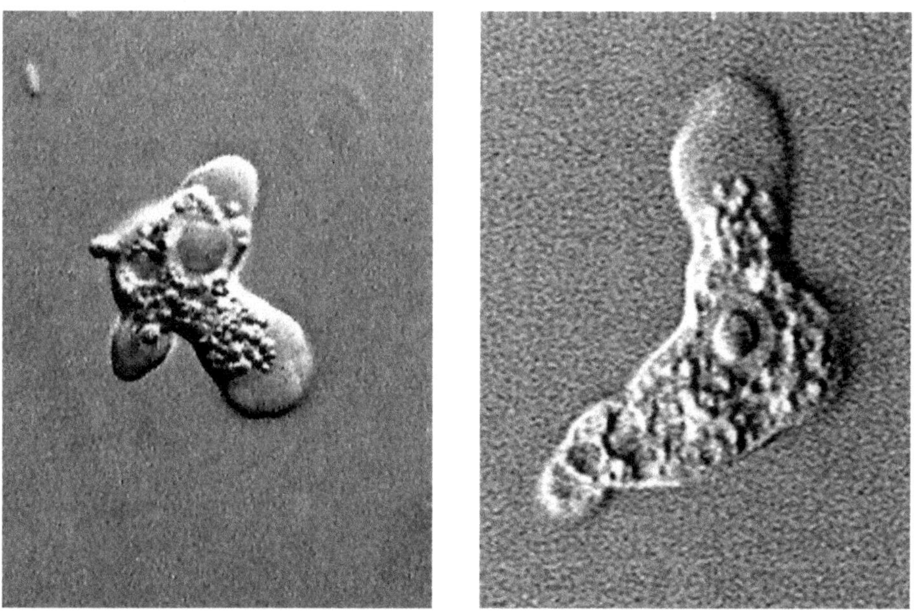

Figure 2.2. *N. fowleri* trophozoite viewed using phase contrast microscopy. X1000. (courtesy: G. S. Visvesvara, CDC, USA).

too high to support the amoebic growth, and Page's amoeba saline (Page, 1985) is used instead.

Samples (cerebrospinal fluid, brain tissue) should be collected aseptically and must not be frozen. For best results, culture should be attempted as soon as the sample arrives or within 24 h of sample collection. If samples are cultured within eight hours of collection, they should be maintained at room temperature. The samples may be refrigerated (2 - 8°C) for longer period of storage, if necessary. Culture growth of *N. fowleri* in axenic condition (i.e. free-from other organisms) has been achieved. This may prove important for identification of pathogenicity or virulence factors, or potential drug targets through the use of next generation sequencing approaches. Since culture procedure does not have a good diagnostic value, no further discussion of culture is provided here.

2.3.4. Enflagellation experiment

Unlike *Acanthamoeba* spp. or *Balamuthia mandrillaris*, the trophozoites of *N. fowleri* convert into flagellates upon certain environmental or experimental conditions in the laboratory. Therefore, this test can be used to differentiate *N. fowleri* from other pathogenic free-living amoebae. The test can be done by mixing one drop of sedimented cerebrospinal fluid containing trophozoites of *N. fowleri* and one mL of distilled water in a sterile tube. After gentle shaking of the tube, one drop of the suspension is transferred onto the center of a coverslip whose edge may be coated with petroleum jelly to prevent dehydration. The slide is then placed in a moist chamber and incubated for 1 - 2 h with periodic checking for free-swimming flagellates. If the sample contains *N. fowleri*, some of the trophozoites (up to 50%) may undergo transformation into pear-shaped biflagellate forms (Figure 2.3). However, one caveat is that not all *N. fowleri* trophozoites can undergo this transformation limiting its use as a confirmatory test (De Jonckheere, 2001).

2.4. Serologic tests

Serologic tests are available but are often not very useful for diagnostic purposes because most primary amoebic meningoencephalitis patients die within a few days of infection; as a result, there is insufficient time to mount a detectable immune response. Previous attempts to detect an antibody response to *N. fowleri* using an immunofluorescence antibody have been mostly unsuccessful (Kilvington and White, 1986; Schuster and Visvesvara, 2004). However, Seidel et al., (1982) reported a specific antibody response to *N. fowleri* in a Californian primary amoebic meningoencephalitis survivor with a titer of 1:4096 demonstrated by immunofluorescence antibody in the serum samples collected as early as 7 days after admission to hospital. The high antibody titers persisted >4 years. Based on immunoblot studies at the U.S. Centers for Disease Control and Prevention, IgM was found to be the principal class of antibody generated by this patient, as well as by three others who developed primary amoebic meningoencephalitis. Moreover, this study also indicated that sera

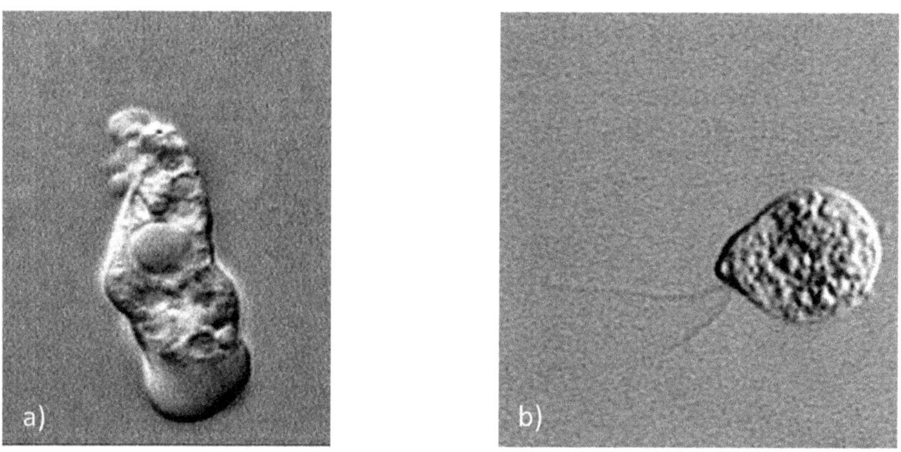

Figure 2.3. *N. fowleri* (a) trophozoite stage, and (b) flagellate stage. X1000. (courtesy: G. S. Visvesvara, CDC, USA).

collected from several individuals with a history of extensive swimming in freshwater lakes in the southeastern United States as well as in California also exhibited IgM antibodies to *N. fowleri*. It is unclear whether these antibodies have any protective role against *N. fowleri* infection (Visvesvara, 2007). A previous study showed that sera from individuals residing in the southeastern United States had significantly greater agglutinating ability (IgM class) than did sera obtained from a northeastern state (Marciano-Cabral et al., 1987). However, it remains unclear whether these antibodies are really an indication of previous exposure or just cross-reactivity with the *N. fowleri* antigens.

In conclusion, serologic tests are not suitable for diagnosis of primary amoebic meningoencephalitis cases because: (a) disease progression in primary amoebic meningoencephalitis cases is so rapid that they may not mount an antibody response, (b) even if there is an antibody response (assuming specific against *N. fowleri*), serology will not be able to distinguish between an acute infection from a past infection, and (c) the test is time consuming.

2.5. Antigen detection tests

Additionally, immunofluorescence assays are shown to be useful for sensitive and specific detection of *N. fowleri* (Anderson and Jamieson, 1972; De Jonckheere et al., 1974). Visvesvara et al., (1987) produced monoclonal antibodies mAbs to *N. fowleri* and used them as probes to identify amoebae in the brain sections of patients who died of primary amoebic meningoencephalitis. These mAbs were characterized for their specificity by the indirect immunofluorescence assay, dot immunobinding assay, and enzyme-linked immunotransfer blot technique. The mAbs reacted intensely with all strains of *N. fowleri* tested originating from different geographic areas in the indirect immunofluorescence assay, and dot immunobinding assay, but showed no reactivity with four other species of *Naegleria*, *N.*

gruberi, *N. jadini*, *N. lovaniensis*, and *N. australiensis*, or a strain of *Acanthamoeba castellanii*. In the enzyme-linked immunotransfer blot technique assay the mAbs reacted with the antigens of *N. fowleri* and produced intensely staining bands at the 160-, 104-, 93-, and 66 kDa regions and several minor bands at the 30- and 50-kDa regions. The mAbs also reacted with the antigens of *N. lovaniensis* and produced a darkly staining band at 160 kDa and a diffusely staining band at 116 kDa, indicating that these antigens were shared by the two species. The mAbs, however, showed no reactivity with *N. jadini* and *N. gruberi* in the enzyme-linked immunotransfer blot technique assay. Later, monoclonal antibodies were produced that showed reactivity with all *N. fowleri* strains tested and with the three morphological forms of *N. fowleri* (trophozoites, cysts, and flagellates) by enzyme-linked immunosorbent assay (Sparagano et al., 1993) and using flow cytometry (Flores et al., 1990). With the help of these *N. fowleri* species-specific mAbs and the IIF test, small numbers of *N. fowleri* amoebae concentrated directly from the contaminated water can be identified quickly, suggesting good sensitivity of the test.

In 1990, Flores and colleagues produced four mAbs against *N. fowleri* that did not react with *A. castellanii* or *A. polyphaga* or other, non-pathogenic species of *Naegleria* (Flores et al., 1990). These mAbs to *N. fowleri* were analyzed by enzyme-linked immunosorbent assay, indirect immunofluorescence assay, and fluorescence flow cytometry to assess specificity and cross-reactivity with axenically cultured *N. fowleri* and *Acanthamoeba* spp. Four mAbs were found to be specific for *N. fowleri* and had no reactivity to *A. polyphaga*. These mAbs showed potential in identifying *N. fowleri* trophozoites in clinical specimens such as brain sections of primary amoebic meningoencephalitis patients in indirect immunofluorescence assays. However, the specificities of the mAbs were not cross checked against other *Naegleria* species such as *N. gruberi*, *N. jadini*, *N. lovaniensis*, and *N. australiensis*, which remains a limitation of these mAbs. Several enzyme-linked immunosorbent assay tests based on mAbs against *N. fowleri* have been developed to detect the amoebic antigens in environmental water samples (Réveiller et al., 2000; 2003; Sparagano et al., 1993), but these were not tested for use with clinical specimens from primary amoebic meningo-encephalitis patients.

Currently the diagnosis of primary amoebic meningo-encephalitis is dependent on recognition of the amoebae in the cerebrospinal fluid, which requires an expert microscopist with adequate training and experience to distinguish amoebae from other host cells. Moreover, it is possible that in some cases, amoebic antigens rather than intact amoebae may be present in the cerebrospinal fluid due to delay in sample processing or due to storage of cerebrospinal fluid in an unfavorable temperature. In such cases, a test that can detect amoebic antigens in the absence of intact amoebae in the cerebrospinal fluid would greatly facilitate early detection of the disease. This would facilitate appropriate therapy early in the infection and increase patient survival.

An indirect immunofluorescence assay for the detection of *N. fowleri* antigen in paraffin-embedded brain tissue slide is routinely performed at Centers for Disease Control and Prevention. The indirect immunofluorescence assay protocol is provided below:

Equipment. Incubator (35 - 39° C), incubation chamber, fluorescent microscope, and pipettes.

Reagents and Media. Xylene, 50% Ethanol, 100% Ethanol, 95% Ethanol, 70% Ethanol, 1X PBS (0.01M. pH 7.6), primary serum -Rabbit anti-*Naegleria* serum, conjugate - FITC labeled Goat anti-rabbit IgG (Sigma F6005 or equivalent), Evans blue counterstain, PVA/DABCO mounting medium, positive and negative control slides.

Supplies, other materials. Pipette tips, microscope slides, cover slips, disposable pipettes, pap pens, forceps, personal protective equipment (gloves, lab coat, safety glasses, etc.).

Test procedure

- Deparaffinization: Place the test slides in xylene for more than 1 hour and less than 24 hours (typically overnight beginning on day of specimen receipt).
- Remove slides from glass soaking chamber and drain excess liquid from slide.
- Transfer the slides to 50% xylene and 50% ETOH for 10 minutes.
- Repeat step 2 and transfer the slides to 100% ETOH for 10 minutes.
- Repeat step 2 and transfer the slides to 95% ETOH for 10 minutes.
- Repeat step 2 and transfer the slides to 70% ETOH for 10 minutes.
- Repeat step 2 and transfer the slides to 1X PBS for 10 minutes.
- Wash the slides 3 times for 5 minutes each with 1X PBS.
- Draw a circle around the tissue section with pap pen. Pap Pens are used for immunohistochemical applications. The hydrophobic properties allow the user to draw barriers on the slide to confine the flow of reagents to a defined area. This prevents the serum from overflowing to the rest of the slide.
- Drain excess PBS and add 1:200 Primary Serum in 1X PBS to the section in sufficient volume to thoroughly cover section.
- Transfer the slides to an incubation chamber in an appropriate container and incubate for 30 minutes at 35-39°C.
- Remove slides from incubation chamber and wash slides 3 times for 5 minutes each with 1X PBS.

- Cover tissue with 1:100 secondary conjugate (Sigma, Goat Anti-Rabbit IgG F6005, or equivalent) in 1X PBS with 3 μL/mL Evans Blue. Return to incubation chamber for 30 minutes at 35-39 °C.
- Remove slides from incubation chamber and wash 3 times for 5 minutes each with 1X PBS.
- Add PVA/DABCO mounting medium and cover slip. Gently press on coverslip with forceps to remove any air bubbles and transfer to a slide folder.

Interpretation of Results. Slides are read using a fluorescent microscope. Positive samples will contain amoebae fluorescing bright green from the FITC-labeled Conjugate/Primary Antibody-Antigen complex. In order to constitute a true positive result, this fluorescence should make the amoeba (trophozoites) visible against the dull red-brown counterstain. Positive and negative controls must be within acceptable limits.

2.6. Molecular Detection

Since microscopy is not always helpful in identifying free-living amoebae to species level, biochemical techniques (such as isoenzyme analysis) have been developed for the specific identification of *N. fowleri* amoebae in culture from the cerebrospinal fluid and brain specimens of patients as well as from the environment (water and soil) (De Jonckheere, 1982; Moss et al., 1988; Visvesvara and Healy, 1980). Recently, molecular techniques such as PCR, nested PCR, and real-time PCR assays have been developed for the specific identification of *N. fowleri* in clinical samples, in cultured amoebae from patients and the environment (Sparagano, 1993; Cogo et al., 2004; De Jonckheere, 2004; Marciano-Cabral et al., 2003; Réveiller et al., 2002; Zhou et al., 2003; Robinson et al., (2006; Behets et al., 2006; Madarová et al., 2009). Both repetitive mitochondrial DNA (Hara and Fukuma, 2005; McLaughlin et al., 1991), as well as genomic DNA (Réveiller et al., 2002; Anonymous, 2012;

Budge et al., 2013; Cetin and Blackall, 2012; Lopez et al., 2012; Madarova et al., 2010; Phu et al., 2013; Qvarnstrom et al., 2006; Schild et al., 2007; Su et al., 2013) have been used to develop these PCR assays. These PCR assays displayed high sensitivity (Table 2.3). A TaqMan real-time PCR based on the 18S small subunit ribosomal RNA (rRNA) gene has been developed (Qvarnstrom et al., 2006; Le Calvez et al., 2012). For cerebrospinal fluid and fresh or frozen brain tissue samples, it takes about 2 - 3 h to identify *N. fowleri* from the time the specimen is received in the laboratory. The detection time is considerably longer for paraffin-embedded tissue samples as they need additional hours to deparaffinize the sample prior to DNA isolation. This TaqMan real-time PCR assay is highly sensitive (as the target, 18S small subunit (SSU) rDNA has an estimated copy number of several hundreds) and can detect a single amoeba in a patient sample. The PCR is also highly specific for *N. fowleri* as it does not amplify other pathogenic free-living amoeba DNA from *A. castellanii* or *B. mandrillaris*. Another advantage of this PCR is that it can detect all the available genotypes of *N. fowleri*. Additionally, conventional PCR, nested PCR and other types of real-time PCR have been developed for detection of *N. fowleri*, but they have not been tested in clinical samples (Table 2.3).

2.7. Conclusions and future work

It is recommended that patients presenting with the clinical picture of bacterial meningitis, but negative cerebrospinal fluid Gram stain and a recent history of nasal freshwater exposure (swimming in warm freshwater, neti pot use, or ritual nasal rinsing) have their cerebrospinal fluid examined for motile trophozoites. Even if the live or motile amoebae are not seen, the cerebrospinal fluid should be subject to amoeba identification using improved molecular detection tools such as real-time PCR.

Table 2.3. PCR based detection of *N. fowleri*.

PCR type	Target	Primer name	Primer sequences (5' to 3')	Probe name	Probe sequence (5' to 3')	Size (bp)	Detection limit	References
Conventional PCR	Repetitive mtDNA	Forward primer	CGTATCTAGTAGATAGAACA	N/A	N/A	300	ND	Hara and Fukuma 2005; McLaughlin et al, 1991
		Reverse primer	CGTAACGACACAAACCTACAGA	N/A	N/A			
Nested PCR	Mp2Cl5	Mp2Cl5.for	TCTAGAGATCCAACCAATGG	N/A	N/A	166	5 pg of DNA or 5 amoebae (for the nested PCR)	Réveiller et al, 2002
		Mp2Cl5.rev	ATTCTATTCACTCCACAATCC	N/A	N/A			
		Mp2Cl5.for-in	GTACATTGTTTTTATTAATTTCC	N/A	N/A	110		
		Mp2Cl5.rev-in	GTCTTTGTGAAAACATCACC	N/A	N/A			
SYBR Green	18S SSU rRNA gene	Nae3-For	CAAACACCGTTATGACAGGG	N/A	N/A	183	DNA from a single amoeba	Schild et al, 2007; Su et al, 2013
		Nae3-Rev	CTGGTTTCCCTCACCTTACG	N/A	N/A			
TaqMan	18S SSU rRNA gene	NaeglF192	GTGCTGAAACCTAG TATTGTAACTCAGT	HEX-labeled probe NfowlP	HEX-AT AGC AAT ATA TTC AGG GGA GCT GGG C-BHQ1	153	DNA from a single amoeba	Qvarnstrom et al, 2006; Budge et al, 2013; Annonymous, 2013; Phu et al, 2013; Cetin and Blackall 201;, Lopez et al, 2012
		NaeglR344	CACTAGAAAAAGCAAACCTGAAAGG					
LightCycler (LC)	Mp2Cl5	Forward primer	TCTAGAGATCCAACCAATGG	Fluorescein probe	CAA GAT CAC TTG TTG AAG GCT GTC-FL	ND	1 copy of target DNA	Madarova et al, 2010
		Reverse primer	GTCTTTGTGAAAACATCACC	LC Red 640 probe	LC Red 640 - CAA ACT CTT TGG CCT CTA TTC CTC TT			

mtDNA = mitochondrial DNA; N/A = Not applicable; ND = Not determined.

Since the route of entry of amoebae is through the nasal cavity, organisms might be identified in the nasal passage or exudates early in the infection. Nasal exudates could be collected and examined for *N. fowleri* by microscopic examination or by molecular detection. This may be achieved using the device proposed by Baig and Khan (2015) which uses a non-ball valve tip attached to a suction syringe to collect nasal exudates.

Utilizing the *N. fowleri* specific mAbs, a lateral flow point-of-care (POC) test may be developed for the diagnosis of primary amoebic meningoencephalitis. This will have several advantages over the existing methods. First, the POC test will not require any expertise in amoeba morphology, which is essential to identify amoebae in the cerebrospinal fluid microscopically. Second, the POC test will be much faster than the conventional or real-time PCR, which would be extremely useful for quick institution of anti-amoebic therapy. Third, the POC test can be performed at the bedside of the patient without requiring any sophisticated instruments. Fourth, the POC test might be less expensive than other tests.

Single Chain Fragment Variable (scFv) proteins are antibody-like molecules that can be expressed on the surface of *Saccharomyces cerevisae*. These molecules have been used in place of traditional mAbs in enzyme-linked immunosorbent assasys without compromising the sensitivity and the specificity of the assay. They are quicker to produce and much cheaper than mAbs (Gray et al., 2012; Grewal et al., 2013; 2014). An *N. fowleri* enzyme-linked immunosorbent assay or POC test based on specific scFv proteins could be useful.

Due to the high price of sophisticated instruments and the expertise needed to perform real-time PCR, this procedure is often unavailable in resource-poor countries. Mahittikorn et al., (2015) developed a novel loop-mediated isothermal amplification assay for *N. fowleri*. Here the amplification products can easily be detected visually using hydroxy naphthol blue within 90 min from the receipt of clinical samples such as cerebrospinal fluid. The loop-mediated isothermal amplification assay appears to be highly specific for *N. fowleri* species; the loop-mediated isothermal amplification assay appears to be able to detect a single trophozoite of *N. fowleri* when used on water and cerebrospinal fluid samples spiked with amoeba trophozoites. The loop-mediated isothermal amplification assay may be an alternative to conventional or real-time PCRs if the assay protocols are appropriately validated against a gold standard assay such as the TaqMan real-time PCR described by Qvarnstrom et al., (2006).

Finally, it is important to note that the causative agents of many central nervous system infections remain unidentified. Some of these infections could be due to *N. fowleri* or to a closely related species. In these unsolved cases, next generation sequencing approaches may be useful to understand the etiology of these infections.

3

Chemotherapeutic and Disinfection Strategies

Abstract

While it is fortunate that primary amoebic meningo-encephalitis is a rare disease, its rarity makes rigorous studies to find the most effective treatment difficult. The majority of studies rely on *in vitro* laboratory testing, mouse models of primary amoebic meningoencephalitis, and the few case reports of primary amoebic meningoencephalitis survivors to inform treatment recommendations for primary amoebic meningoencephalitis. In addition, the acquisition of drug resistance is a constant threat due to challenges in developing novel drugs. This chapter will present current treatment recommendations based on recent survivor case reports followed by a discussion of the drugs, treatment interventions used, promising new therapies and disinfection strategies.

3.1. Current Treatment Recommendations

Although most cases of primary amoebic meningo-encephalitis caused by *N. fowleri* infection have been fatal, there have been five well-documented survivors (one from Australia in 1971, one from the United States in 1978, one from Mexico in 2003, and two from the United States in 2013) who received a combination of drug treatments, all of which included amphotericin B (Table 3.1) (Anderson and Jamieson, 1972; Seidel et al., 1982; Vargas-Zepeda et al., 2005; Siddiqui and Khan, 2012; Linam et al., 2015; Capewell et al., 2014; Cope et al., 2016). Based on these past case reports and in light of the recent primary amoebic meningo-encephalitis survivors in the United States in 2013, the U.S. Centers for Disease Control and Prevention (CDC) currently recommends that any patient suspected to have primary amoebic meningoencephalitis should be started on the following combination of drugs: deoxycholate amphotericin B intravenously (IV) and intrathecally (IT), an azole drug such as fluconazole IV or orally (PO), azithromycin IV or PO, rifampin IV or PO, miltefosine PO, and dexamethasone IV (Table 3.2).

Table 3.1. Drug Regimens Used to Treat Five Well-Documented Survivors of Primary Amebic Meningoencephalitis.

Australian Survivor (1971)	U.S. Survivor (1978)	Mexico Survivor (2003)	U.S. Female Survivor (2013)	U.S. Male Survivor (2013)
Amphotericin B (IV, intrathecal, and via ventricular reservoir)	Amphotericin B (IV and intrathecal)	Amphotericin B (IV)	Amphotericin B (IV and intrathecal)	Amphotericin B (IV and intrathecal)
	Rifampin (oral)	Rifampin (oral)	Rifampin (IV/oral)	Rifampin (oral)
	Miconazole (IV and intrathecal	Fluconazole (IV and oral)	Fluconazole (IV/oral)	Fluconazole (IV)
	Dexamethasone	Dexamethasone (IV)	Dexamethasone (IV)	Dexamethasone (IV)
	Sulfisoxazole (IV) – discontinued after PAM diagnosed	Ceftriaxone (IV)	Azithromycin (IV/ oral)	Azithromycin (oral)
			Miltefosine (oral)	Miltefosine (oral)

3.1.1. Amphotericin B

Naegleria spp. are highly sensitive to the antifungal drug amphotericin B making it the mainstay of primary amoebic meningoencephalitis treatment. All of the well-documented survivors have been treated with amphotericin B (Table 3.1). Minimum amoebicidal concentrations of amphotericin B were determined to be 0.026 - 0.078 mg per mL for three different clinical isolates of *N. fowleri* tested *in vitro* (Duma et al., 1971). Many other strains tested showed similarly low, though not necessarily identical, amoebicidal drug levels. Ultrastructural examination of amoebae treated with amphotericin B revealed membrane distortions, including the nuclear envelope, rough and smooth endoplasmic reticula, and plasma membrane blebbing (Schuster and Rechthand, 1975).

Amphotericin B was identified as a candidate for primary amoebic meningoencephalitis treatment soon after the first report of *Naegleria* causing fatal meningoencephalitis in 1965. It was included in early *in vitro* studies because of its success in the treatment of fungal infections of the central nervous system as well as its demonstrated effect for the treatment of leishmaniasis (Carter, 1969). In Carter's findings, amphotericin B was found to have a minimum inhibitory

Table 3.2. Recommended Treatment for Primary Amebic Meningoencephalitis Caused by *Naegleria fowleri*.

Drug	Dose	Route	Maximum Dose	Duration	Comments
Amphotericin B	1.5 mg/kg/day in 2 divided doses	IV	1.5 mg/kg/day	3 days	
then	1 mg/kg/day once daily	IV		11 days	14-day course
Amphotericin B	1.5 mg once daily	Intrathecal	1.5 mg/day	2 days	
then	1 mg/day every other day	Intrathecal		8 days	10-day course
Azithromycin	10 mg/kg/day once daily	IV/PO	500 mg/day	28 days	
Fluconazole	10 mg/kg/day once daily	IV/PO	600 mg/day	28 days	
Rifampin	10 mg/kg/day once daily	IV/PO	600 mg/day	28 days	
Miltefosine	Weight<45 kg 50 mg BID Weight>45kg 50 mg TID	PO	2.5 mg/kg/day	28 days	50 mg tablets
Dexamethasone	0.6 mg/kg/day in 4 divided doses	IV	0.6 mg/kg/day	4 days	

concentration (MIC) lower than that required to inhibit *Cryptococcus neoformans*, indicating that *Naegleria* was highly sensitive to amphotericin B. Additionally, the MIC was lower than the level attainable by intravenous therapy in human cerebrospinal fluid (Carter, 1969). Finally, Carter also demonstrated amphotericin B had a protective effect on mice inoculated with *N. fowleri*. Subsequent *in vitro* studies confirmed these findings and provided further support for the use of amphotericin B as first line therapy for primary amoebic meningoencephalitis (Duma et al., 1971; Schuster and Rechthand, 1975; Duma, 1970).

While amphotericin B looked promising in laboratory studies, its clinical use was limited by its toxicities, including acute infusion-related reactions and dose-related nephrotoxicity and lack of survival in most of the primary amoebic meningoencephalitis cases in which it was used. In cases in which amphotericin B was used unsuccessfully, the patient's death was assumed to be either due to the inevitable outcome of the disease, a delay in initiating therapy, or both of these factors rather than because of drug resistance to amphotericin B. Studies have shown that *Naegleria* isolates from deceased primary amoebic meningoencephalitis patients have sensitivity to amphotericin B when tested *in vitro* (Duma and Finley, 1976; Tiewcharoen et al., 2002).

Since the initial studies of amphotericin B and its use for primary amoebic meningoencephalitis treatment, new lipid-associated formulations of amphotericin B (amphotericin B lipid complex, liposomal amphotericin B, and amphotericin B colloidal dispersion) have been introduced with improved toxicity profiles. However, when deoxycholate amphotericin B was compared with liposomal amphotericin against *N.fowleri*, the MIC for deoxycholate amphotericin B was 0.1 μg/mL, while that of liposomal amphotericin was 10 times higher at 1 μg/mL. Liposomal amphotericin was less effective in the mouse model and in *in vitro* testing than the more toxic form of deoxycholate amphotericin (Goswick and Brenner, 2003; 2003a). Amphotericin methyl ester was also less effective in the mouse model (Ferrante 1982; Lee and Karr, 1979). Thus, because of the extremely poor prognosis of primary amoebic meningoencephalitis caused by *N. fowleri*, healthcare providers might consider using deoxycholate amphotericin B instead of the liposomal or lipid complex formulation. However, if deoxycholate amphotericin B is not immediately available, treatment should be initiated with a non-deoxycholate formulation to facilitate prompt treatment of the patient.

3.1.2. The Azoles

Another class of antifungal drugs, the azoles (including clotrimazole, miconazole, ketoconazole, fluconazole, and voriconazole) have been tested against *Naegleria* isolates *in vitro* and in mouse models and have also been used to treat primary amoebic meningoencephalitis survivors. The toxicities of amphotericin B spurred research into other drug candidates and several studies found members of the azole class drugs to have activity against *N. fowleri*, although none as active as amphotericin B (Duma and Finley 1976). Newer azole agents have also shown effective inhibitory activity against *N. fowleri* (Tiewcharoen et al., 2002). Tiewcharoen et al. (2002) found ketoconazole to have the lowest MIC of the azoles tested (ketoconazole, fluconazole, and itraconazole). Voriconazole has inhibitory and amoebicidal effects on *N. fowleri* (Schuster et al., 2006). Among the well-documented survivors, miconazole and fluconazole have been used as part of successful treatment regimens (Seidel et al., 1982; Vargas-Zepeda et al., 2005; Linam et al., 2015; Capewell et al., 2014).

3.1.3. Macrolides

Continued investigation for drugs with activity against *N. fowleri* resulted in the finding that azithromycin showed promise as a useful agent for the treatment of primary amoebic meningoencephalitis. Azithromycin was investigated because of reports describing *in vitro* sensitivity against *Acanthamoeba* spp. (another free-living amoeba) and its activity in experimental toxoplasmosis (Araujo et al., 1991; Schuster and Visvesvara, 1998). Schuster and colleagues tested three macrolide antibiotics against *N.fowleri*, finding erythromycin to have only minimal inhibitory effect on the growth of amoebae while azithromycin was inhibitory at 1 - 5 µg per mL and amoebicidal at 10 µg per mL with clarithromycin activity falling in between these two (Schuster and Visvesvara, 2001). Goswick and Brenner (2003) demonstrated that azithromycin has *in vitro* activity against *N. fowleri* and protected 100% of mice in an experimental primary amoebic meningoencephalitis model, despite having an MIC approximately 123 times greater than amphotericin B. They postulated that azithromycin's success in the mouse model might be due to its unique pharmacokinetics, including a long elimination half-life and high tissue levels, including good penetration of the blood-brain barrier. Additional work by this group demonstrated that amphotericin B and azithromycin are synergistic against *N. fowleri*, suggesting that the combined use of these two drugs might be a successful strategy in the treatment of primary amoebic meningoencephalitis (Soltow and Brenner, 2007). Azithromycin was used as part of the treatment regimen for two recent survivors in the United States (Table 3.1) (Linam et al., 2015; Cope et al., 2016). Clarithromycin, another commonly used macrolide antibiotic, does not have *in vitro* activity against *N. fowleri*, although it has been used in the treatment of other free-living amoeba infections (Kim et al., 2008).

3.1.4.Rifamycins

In 1977, Thong et al. first reported that rifamycin suppressed the growth of *N. fowleri* in culture (Thong et al., 1977). Further study of rifamycin given alone in a primary amoebic meningoencephalitis mouse model did not result in any survivors; however, amphotericin B and rifamycin given together in a mouse model showed a synergistic effect (Thong et al., 1979). Other studies have not shown rifamycin drugs to be effective (Goswick and Brenner, 2003; Stevens et al., 1981). Rifampin has been included in the treatment regimen of four well-documented survivors (Table 3.1). While the *in vitro* and mouse model data have not supported the use of rifampin for the treatment of primary amoebic meningoencephalitis, its inclusion in the drug regimen given to most of the well-documented survivors has resulted in it being continually recommended for primary amoebic meningoencephalitis treatment.

3.1.5. Rokitamycin

Kim et al., (2008) examined the amoebicidal activities of antibacterial agents including clarithromycin, erythromycin, hygromycin B, neomycin, rokitamycin, roxithromycin and zeocin. The results showed that the growth of amoeba was effectively inhibited by treatment with hygromycin B, rokitamycin and roxithromycin. Notably, when *N. fowleri* trophozoites were treated with rokitamycin, the minimal inhibitory concentration was 6.25 microg/mL on Day 2. In the treatment of experimental meningoencephalitis due to *N. fowleri*, survival rates of mice treated with roxithromycin and rokitamycin were 25% and 80%, respectively, over 1 month. The mean time to death for roxithromycin and rokitamycin treatment was 16.2 days and 16.8 days, respectively, compared with 11.2 days for control mice. Finally, rokitamycin showed both *in vitro* and *in vivo* therapeutic efficacy against *N. fowleri* and may be a candidate drug for the treatment of primary amoebic meningoencephalitis (Kim et al., 2008).

3.1.6. Tetracyclines

Tetracycline was first reported to have *in vitro* activity against *Naegleria fowleri* in 1977 (Thong et al., 1977). It was shown subsequently to have marked synergism with amphotericin B in a mouse model (Thong et al., 1978; 1979). Further study demonstrated that minocycline was also synergistic with amphotericin B, in addition to tetracycline (Lee et al., 1979). Gogate et al. (1986) found treatment of a mouse model with tetracycline to be ineffective (mortality 90%) but when it was used in combination with amphotericin B or rifampicin, mortality dropped to 20-30%. None of the five well-documented human primary amoebic meningoenceph-alitis survivors were treated with a tetracycline drug (Table 3.1).

3.1.7. Miltefosine

Miltefosine was originally an anticancer agent that was then identified and studied as a treatment for leishmaniasis (Marschner et al., 1992). Because of its antiparasitic activity and known ability to cross the blood-brain barrier and concentrate in brain tissue, miltefosine was tested against *N. fowleri* and found that it had amoebastatic activity with amoebicidal activity at higher concentrations (Schuster et al., 2006; Marschner et al., 1992). An additional study again demonstrated miltefosine's *in vitro* activity against *N. fowleri* as well as improved survival in a primary amoebic meningoencephalitis mouse model (Kim et al., 2008). Because of miltefosine's potential for primary amoebic meningoencephalitis treatment as well as demonstrated activity against other free-living amoebae (*Acanthamoeba* and *Balamuthia*), the U.S. CDC began recommending and facilitating its use in the United States under single-patient emergency Investigational New Drug (IND) protocols. It was first given to a primary amoebic meningoencephalitis patient in 2010 in combination with amphotericin B, azithromycin, fluconazole, and rifampin; however, the drug was not administered to the patient until 5 days after the patient's initial presentation to the hospital and the patient did not survive (CDC unpublished data). Recently, the U.S. CDC has been able to procure a supply of miltefosine under an expanded access IND allowing for rapid deployment to primary amoebic meningoencephalitis patients in the U.S (CDC, 2013). In 2013, two U.S. primary amoebic meningo-encephalitis patients received miltefosine from CDC allowing for its administration (in combination with amphotericin B, azithromycin, fluconazole, and rifampin, see Table 3.1) shortly after their hospital admissions, possibly contributing to their survival (Linam et al. 2015; Cope et al., 2016). Outside of the U.S., miltefosine might be more readily available and should be included, if at all possible, in a primary amoebic meningoencephalitis treatment regimen.

3.2. Strategies to Reduce Elevated Intracranial Pressure

The common pathway leading to the death of primary amoebic meningoencephalitis patients is overwhelming inflammation of the brain with resulting cerebral oedema leading to elevated intracranial pressure and subsequent brain herniation. Therefore, the successful treatment of a primary amoebic meningoencephalitis patient must focus on managing this cascade of events in addition to killing the amoeba itself with the drugs discussed previously. Four of the well-documented survivors received dexamethasone as part of their treatment to control the inflammatory process, thereby reducing intracranial pressure (Table 3.1). The most recent well-documented survivor who recovered with no neurologic sequelae was managed aggressively for elevated intracranial pressure. Management of this individual included drainage of cerebrospinal fluid via an external ventricular drain, hyperosmolar therapy with mannitol and 3% saline, moderate hyperventilation (goal P_aCO_2: 30 - 35 mm Hg), and induced hypothermia (32 - 34°C) (Linam et al., 2015). Regardless of what modalities are used, primary amoebic meningoencephalitis patients should have their intracranial pressure closely monitored likely necessitating the involvement of a neurosurgeon in the care of the patient, if at all possible.

3.3. Other Drugs

Other drugs have shown *in vitro* activity against *N. fowleri* but have not been used as part of the treatment regimen for any of the well-documented survivors. Antipsychotic drugs of the phenothiazine group including trifluoperazine and chlorpromazine have been shown to have *in vitro* activity as well as improved survival in a primary amoebic meningoencephalitis mouse model (Kim et al., 2008; Schuster and Mandel, 1984; Ondarza et al., 2006). A metabolic product of cyclophosphamide, acrolein, inhibited growth and enflagellation of *N. fowleri*. Acrolein at 40 microM was amoebicidal. Acrolein injured starved cells and amoebae at 5°C and growing *N. fowleri* (Zhang et al., 1988). Several studies have investigated whether or not various antimalarial drugs have any activity against *N. fowleri*. Chloroquine and pyrimethamine halve the multiplication rate of *N. fowleri* trophozoites (Dhu 1982). However, pyrimethamine had no effect on mortality of mice infected with *N. fowleri* (Thong 1978). Artemisinin and its derivatives were shown to have some *in vitro* activity against *N. fowleri* but *in vivo* experiments in mice showed only slight improvement in mean survival time (Gupta et al., 1995; Cooke et al., 1987). Cannabinoids have also been shown to inhibit the growth of *N. fowleri* (Pringle et al., 1979).

The efficacies of several tricyclic neuroleptics, antimycotics and antibiotics with antiproliferative activities were investigated against *N. fowleri*. The most effective drugs against *N. fowleri* expressed as (IC50) were as follows: the antimycotics ketoconazole and amphotericin B, followed by trifluoperazine, mepacrine, chlorpromazine, miconazole, and metronidazole. The least effectives were rifampicin and pentamidine. The most potent growth inhibitors (MIC100) against *N. fowleri* were the antimycotics amphotericin B and ketoconazole and the neuroleptic trifluoperazine (Ondarza et al., 2007).

Recently, automated, high throughput screening methodology has been used to identify new compounds with potential activity against *N. fowleri*. Corifungin, a water-soluble polyene macrolide, is one compound identified using this methodology. Using transmission electron microscopy, *N. fowleri* trophozoites incubated with corifungin showed disruption of cytoplasmic and plasma membranes, alterations in mitochondria, and complete lysis of amoebae (Debnath et al., 2012). Additionally, using a primary amoebic meningoencephalitis mouse model, corifungin resulted in 100% survival of mice for 17 days post-infection. Rice et al. (2015) have also recently used medium and high throughput drug discovery methods to determine that both mono- and di-amidino derivatives have potency against *N. fowleri* and are known to cross the blood-brain barrier, making them attractive candidates for further study in the treatment of primary amoebic meningoencephalitis. Finally, the effect of synthetic antimicrobial peptides on *N. fowleri* showed that tritrpticin caused apoptosis, affected the elasticity of the surface membrane, and reduced the size of the nuclei in *N. fowleri* trophozoites (Tiewcharoen et al., 2014). Notably, tritrpticin was not toxic against SK-N-MC cells.

3.4. Amoebicidal activity of animal serum

The sera of 16 species of wild animals representing 5 classes of vertebrates were assayed for amoebicidal activity against species of *Naegleria*. The greatest activity was observed for sera of bullfrogs, muskrats, and raccoons, all of which are animals associated with water. In contrast, the sera from animals such as toads, box turtles, sparrows, and squirrels exhibited minimal or no amoebicidal activity (John and Smith 1997). Treatment of axenic *N. gruberi* cultures with alligator serum resulted in time-dependent amoebicidal activity, with measurable activity at 5 min and maximal activity occurring at 20 min. The amoebicidal activity was concentration dependent, with measurable activity at 25% serum, whereas treatment of amoebas with undiluted serum resulted in only 16% survival. The amoebicidal effects of alligator serum were temperature dependent, with optimal activity at 15-30°C and a decrease in activity below 15°C and above 30°C. The amoebicidal activity of alligator serum was heat labile and protease sensitive, indicating the proteinaceous nature of the activity, and was also inhibited by ethylenediaminetetraacetic acid, which indicated a requirement for divalent metal ions. These characteristics strongly suggest that the amoebicidal properties of alligator serum are because of complement activity (Merchant et al., 2004).

3.5. Other agents as disinfectants

3.5.1. Chlorine

De Jonckheere et al., (1976) showed the destructive action of chlorine on pathogenic *N. fowleri*. This provided a basis for the destruction of amoebic cysts in drinking water and swimming pools. *Naegleria* cysts were sensitive more to chlorine than to bromine. Notably, the sensitivity to these disinfectants stays the same at 25°C and 35°C. Also, *Naegleria* spp. were more sensitive to chlorine and chlorine dioxide than *Acanthamoeba* spp. (Cursons et al., 1980).

Later studies compared the resistance of *N. gruberi* cysts to chlorine in the presence of cyanuric acid at pH 5 and 7. An amperometric membrane electrode was used to measure HOCl concentrations independently of the chlorinated cyanurate species, thus permitting an analysis of the role of free chlorine versus chlorinated cyanurates in cyst inactivation. In the presence of cyanuric acid, the products of the HOCl residual and the contact time required for 99% cyst inactivation were 8.5 mg. min/liter and 13.9 mg . min/liter at pH 5 and 7, respectively. The Watson's Law coefficients of dilution (n) were 1.3 and 1.6 at pH 5 and 7, respectively. These results suggested that HOCl is the predominant cysticide with no measurable cysticidal effect of the chlorinated cyanurate species (Engel et al., 1983).

Cassells et al., (1995) demonstrated that a combination of silver and copper ions were ineffective at inactivating

Naegleria at 80 and 800 μg per L respectively. However addition of 1.0 mg per L free chlorine produced a synergistic effect, with superior inactivation relative to either chlorine or silver-copper in isolation.

3.5.2. Peracetic acid and monochloramine

Although hypochlorite is the most commonly used biocide to disinfect cooling water systems, however, its negative impact on the aquatic environment, ecologically less harmful alternatives have been sought. Biocidal activity of monochloramine and peracetic acid makes them good candidates for inactivation of pathogenic *Naegleria* species. The biocidal activity of hypochlorite was 8 - 10x stronger than that of the two investigated substances. Hypochlorite, at a concentration of 0.5 mg per L, killed 100% *Naegleria lovaniensis* after 1 h exposure (25°C, pH 7.3 - 7.4). To achieve similar results with monochloramine and peracetic acid, 3.94 mg per L or 5.33 mg per L had to be used, respectively (25°C, pH 8) (Ercken et al., 2003). Later studies tested the efficacy of monochloramine against planktonic forms (trophozoites and cysts) and also biofilm-associated cells of *N. fowleri* as they are often associated with biofilms. The effective range varied from 4 to 17 mg Cl_2 per L at 25°C and pH 8.2 on both planktonic and biofilm associated cells. These findings have impact on water treatment strategies against amoebae when controlling *N. fowleri* in man-made water systems such as cooling towers or hot water systems (Goudot et al., 2014).

3.5.3. Simulated solar disinfection

The antimicrobial activity of simulated solar disinfection in the presence and absence of riboflavin against *N. fowleri* was investigated. Assays were conducted in transparent 12 well microtitre plates containing a suspension of test organisms in the presence or absence of 250 μM riboflavin. Plates were exposed to simulated sunlight at an optical irradiance of 550 Wm^{-2} (watts per square metre) delivered from a SUNTEST™

CPS+ solar simulator. Aliquots of the test suspensions were taken at set time points and the viability of the organism was determined. The exposure to simulated solar disinfection at an optical irradiance of 550 Wm^{-2} for up to 6h resulted in significant inactivation of *Naegleria*.

3.5.4. Pulsed electric fields

The effects of pulsed electric fields on the inactivation of trophozoite form of *Naegleria* showed that amoebae eradication is modulated by pulse parameters, composition of the pulsing medium, and physiological state of the cells. Cell survival is not related to the energy delivered to the cell suspension during the electrical treatment. For a given energy, a strong field applied for a short cumulative pulse duration affects viability more than a weak field with a long cumulative pulsation. It was also determined that the optimal electrical conditions to obtain an inactivation rate higher than 95% while using the least energy. Flow processes allow to treat large-scale volumes. These results show that the most efficient flow process for amoeba eradication requires a field parallel to the flow and that pulsed electric fields offers a new and attractive method for inactivating amoebae in large volumes of fresh water (Vernhes et al., 2002).

3.5.5. Inhibition of *Naegleria fowleri* by microbial iron-chelating agents

Deferrioxamine B and rhodotorulic acid, iron-chelating agents of microbial origin, exerted a pronounced inhibitory effect on pathogenic *N. fowleri* at microgram levels (Newsome and Wilhelm, 1983). This inhibition is diminished by adding iron to the chelators before incubation with *Naegleria* isolates. These and related microbial iron chelators occur naturally in the environment. This could be of considerable ecological significance and provides a novel hypothesis to account for the proliferation of pathogenic *Naegleria* spp. in certain aquatic habitats.

3.5.6. Delta 9-tetrahydrocannabinol

Delta 9-tetrahydrocannabinol (delta 9-THC) inhibited the growth of *N. fowleri* (Pringle et al., 1979). The delta 9-THC is amoebistatic at 5 to 50 µg per mL. The delta 9-THC prevents enflagellation and encystation, but does not impair amoeboid movement. The delta 9-THC prevented the cytopathic effect of *N. fowleri* on African green monkey (Vero) cells and human epithelioma (HEp-2) cells in culture. delta 9-THC afforded modest protection to mice infected with *N. fowleri*.

3.6. Resistance of pathogenic *Naegleria* to some common physical and chemical agents.

Resistance of pathogenic *Naegleria* to drying, low and high temperature, and two halogens has been studied (Chang, 1978). Drying made trophozoites nonviable instantaneously and cysts nonviable in less than 5 min. Trophozoites degenerated in hours at temperatures below 10°C and in minutes when frozen; cysts survived according to the equation th - t0/theta 1,440/1.122T (t0 is survival at 0°C), but 1.5 h at -10°C to 1 h at -30°C. At 51, 55, 58, 63, and 65°C, trophozoites survived about 30, 10, 5, 1 and less than 0.5 min, respectively, cysts survived three to four times longer at 51°C and six to seven times longer at 55 to 65°C. Cyst destruction rates by heat indicated first-order kinetics with 25,400 cal per 1°C for energy of activation (Chang, 1978). Cyst destruction rates by free chlorine and I2 also conformed to first-order kinetics. Concentration-contact time curves yielded concentration coefficient values of 1.05 for free chlorine and 1.4 for I2 and point to superchlorination as an effective means of destroying the cysts if free residuals are used as a guide and allowance is provided for low temperature and/or high pH waters.

3.7. Future Drug targets

Ondarza (2007) proposed following drug targets against *N. fowleri*: a) enzymes which are secreted by amoebae to invade the human host, for example proteinases, phospholipases and pore forming peptides, b) glycolytic enzymes from *Entamoeba* and *Naegleria*, like the PPi-dependent phosphofructokinase that differ from the host enzyme, c) thiols and enzymes of redox metabolism, such as the trypanothione/trypanothione reductase that maintains the reducing environment within the cell, d) antioxidant enzymes to regulate the oxidative stress produced by the phagocytic cells of the host or by the parasite metabolism, like the trypanothione peroxidase in connection with the NADPH-dependent trypanothione/trypanothione reductase which maybe is present in *N. fowleri*, e) enzymes for the synthesis of trypanothione like the ornithine decarboxylase, spermidine synthase and trypanothione synthetase, f) some of the proteins that assemble the secretory vesicles with the cell membrane, like the synaptobrevins and finally, g) encystation pathways and cyst-wall assembly proteins.

3.8. Future prospects: Strategies to deliver antiamoebic drugs

It has recently been proposed that drugs be delivered through the intranasal route, across the cribriform plate (an anatomically porous bone) that is located at the roof of the nasal cavity (Baig and Khan, 2015). Due to the sieve like nature of the cribriform plate, the tips at the terminals of the proposed device with ball-valve function (one way drug delivery) are designed to reach and deliver drugs in vaporized form to area of the brain (the inferior surface of the frontal lobe), a site where *N. fowleri* reaches and concentrates as its natural course of the disease, before spreading to the rest of the central nervous system. The underlying rationale for proposing such a portal is that many therapeutically-desired drugs like amphotericin B fail to attain MIC in the central nervous system, when administered systemically. The proposed route will be advantageous in that it will (i) bypass selectivity of the blood-brain barrier that limits drug permeability to the brain tissue, (ii) trail the natural route of entry of *N. fowleri* into the brain, (iii) target the site of infection, (iv) exert lethal effect of drugs by attaining MIC at a lower dose without venous drainage, (v) avoid adverse effects due to systemic administration. Olfactory targeting through intranasal delivery would provide access to the central nervous system without affecting the integrity of the blood-brain barrier. This can be achieved through the use of a modified intranasal instrument (US Patent- 8146587 B2). It was proposed that this instrument should be modified to extend its nasal terminals so it can reach porous cribriform plate. Additionally, a ball-valve effect should be added to the tip of the modified device so that to deliver the drug in vaporized form and prevent regurgitation to the flow system. The proposed route offers clear benefits over the conventional intravenous route, in that it would enable the use of water soluble drugs to attain MIC at the epicenter of infection, overcome the blood-brain barrier hindrance, minimal adverse effects by enjoying the benefits of topical administration and dose adjustment at a faster pace. Further research is needed to test the effectiveness of the proposed route to administer drugs in the management of primary amoebic meningoencephalitis due to *N. fowleri* and possibly other pathogens causing brain infections.

4

Pathogenesis

Abstract
Primary amoebic meningoencephalitis usually occurs after the inhalation of water containing amoebae or flagellates. It also has been suggested that inhaling cysts, during dusts storms, for example, could lead to infection. Amoeba penetrate the nasal mucosa and the cribiform plate and travel along the olfactory nerves to the brain. Amoebae first invade the olfactory bulbs and then spread to the more posterior regions of the brain. Within the brain they provoke inflammation and cause extensive damage to the tissue. In view of the devastating nature of *N. fowleri* infection and the problems associated with successful prognosis, here we describe current understanding of the pathogenesis of primary amoebic meningoencephalitis, as well as factors that affect virulence of *N. fowleri*, with an eye to identify potential therapeutic targets.

4.1. Axenic growth and pathogenic potential of *N. fowleri*

Axenic cultivation gradually decreases virulence of virulent strains (De Jonckheere, 1979b). The decline in virulence is most rapid during the first 2 years. Virulence could be restored to original levels by serial passages in mice (John and Howard, 1993). Although virulence could be enhanced in low virulent strains by brain passage and passages in Vero cell cultures, but could not be induced by these methods in non-virulent strains isolated from the environment (De Jonckheere, 1979b; Lee et al., 1983; John and John, 1994; Gupta and Das, 1999). However, long term storage in the cyst form retains virulence. Warhurst et al., (1980) showed that *N. fowleri* cysts stored at 4°C for eight months retained virulence. The ability of *N. fowleri* to grow at high temperatures is a useful indicator of virulence (Griffin, 1972).

The composition of axenic growth media also influences growth behaviour and morphology, and pathogenicity of *N. fowleri*. Growth in axenic medium (consisting of 10 g of isoelectric casein, 1.325 g of Na_2HPO_4, 2.5 g of glucose, 0.8 g of KH_2PO_4, 5 g of yeast extract, 900 mL of distilled water, and 100 mL of fetal calf serum) allowed separation of pathogenic from weakly pathogenic *N. fowleri* strains, since only the former showed abundant growth in this medium. Strains showing only moderate growth or no growth at all in this axenic medium were found to be non-pathogenic for mice (De Jonckheere, 1977). Notably, mortality was greater for mice inoculated with amoebae harvested at late logarithmic and early stationary growth phases than it was for amoebae harvested at early logarithmic and late stationary growth phases.

Later studies show that trophozoites maintained in Nelson's medium show rapid *in vitro* proliferation and were pathogenic for mice. In contrast, *N. fowleri* cultured in PYNFH medium exhibited a low pathogenicity and slower growth. However, cultivation of the amoeba in Nelson's medium or PYNFH medium supplemented with liver hydrolysate resulted in trophozoites that were pathogenic in mice. These findings suggest that the presence of liver hydrolysate result in increased proliferation of trophozoites and enhanced pathogenicity of *N. fowleri* in mice (Burri et al., 2012; Zysset-Burri et al., 2014). Although the virulence of axenically cultivated *Entamoeba histolytica* increases following growth with cholesterol, *N. fowleri* incubated with cholesterol for 6 months and tested in mice for changes in virulence showed less virulence for mice than the same strain grown without cholesterol (John and McCutchen, 1995). However, serial passages *in vivo* is effective in enhancing virulence (De Jonckheere, 1979b; Lee et al., 1983; John and Howard, 1993; John and John, 1994; Gupta and Das, 1999).

4.2. *In vivo* models

Susceptible animals were eastern gray squirrel, sheep, hispid cotton rat, muskrat, and house mouse. Mammals tested that were not susceptible at a dose of 10^6 intranasally were opossum, raccoon, and eastern cottontail rabbit. It was suggested that perhaps rodents and humans share a common anatomical or physiological determinant that makes them susceptible to infection with *N. fowleri* (John and Hoppe 1990; Young et al., 1980; Simpson, 1982). For sheep, intranasal inoculation of *N. fowleri* resulted in death in 7 days post-inoculation (Simpson 1982). In these animals, The olfactory lobes were distinctly soft and friable. Histologic findings indicated suppurative leptomeningitis and haemorrhagic necrosis in the olfactory lobes. Amoebae were disseminated in the necrotic areas, particularly in perivascular locations; vasculitis was also observed. Meningitis and perivascular cuffing with lymphocytes were evident in the cerebrum, cerebellum, pons, medulla, and cervical spinal cord. Electron microscopy disclosed trophozoites only in the olfactory lobes. The amoebae contained a central nucleus with a distinct, electron-dense nucleolus. The cytoplasm contained myelinated figures, lipid-like vacuoles, open vesicles, electron-dense granules, mitochondria, numerous free ribosomes, scant rough endoplasmic reticulum, and occasionally a phagocytized erythrocyte. Trophozoites were grouped close to arterioles, except when phagocytized by a neutrophil or endothelial cell.

Later studies show that rabbits are useful models for primary amoebic meningoencephalitis. *N. fowleri* was cultured in a liquid axenic medium, and then injected intracisternally into New Zealand White rabbits. Inocula of 10^3 or 10^5 trophozoites consistently produced a sanguinopurulent meningitis; duration of survival of rabbits was 57 or 45 h, respectively. Counts of cells in cerebrospinal fluid were proportional to the size of inoculum used; white blood cell counts ranged from 30 to 1,055 cells per mm^3, and red blood cell counts from five to 8,640 cells per mm^3. Necropsies revealed severe basilar meningoencephalitis with extensive hemorrhagic necrosis and polymorphonuclear cell infiltration. Trophozoites of *N. fowleri* were seen within the meningeal exudate and the brain parenchyma. Potential applications of this model include studies of the host response to amoebae in the cerebrospinal fluid, and evaluation of the optimal route of administration of other chemotherapeutic agents (Smego and Durack, 1984).

The mouse model has been used most extensively to study the pathogenetic mechanisms, and the role of host immunological mechanisms in resistance and susceptibility to *N. fowleri* infection. Groups of mice were placed in water containing from 10^2 to 10^6 amoebae per mL and allowed to swim for 2.5 to 20 min. Mouse mortality ranged from 0 to 70% and was dependent upon the concentration of amoebae per mL and the length of swimming exposure. That swimming mice can develop fatal *Naegleria* infection confirmed the usefulness of this model for studying primary amoebic meningoencephalitis (John and Nussbaum, 1983). Of note, BALB/c mice are more susceptible than ICR mice. The susceptibility of animals is influenced by the weight (shorter in 15 g mice than in the heavier groups), and age (older mice are less sensitive than younger ones) (Holbrook and Parker, 1979; Ahn and Im 1984). Notably, Ahn and Im, (1984) showed that the susceptibility of BALB/c mice was not influenced by sex. However, Haggerty and John, (1978) showed that the susceptibility of C57BL/6 was influenced by sex as females were significantly more resistant to infection than males.

4.3. *In vitro* models

4.3.1. Organotypic slice cultures from rat brain tissue to study *N. fowleri* infection

Gianinazzi et al., (2005) reported infection of organotypic slice cultures from rat brain with *N. fowleri* and compared the findings in this culture system with *in vivo* infection in a rat model of primary amoebic meningoencephalitis, that proved complementary to that of mice. The findings showed that brain morphology, as present *in vivo*, is well retained in organotypic slice cultures, and that infection time-course including tissue damage parallels the observations *in vivo* in the rat, suggesting that organotypic slice cultures from rat brain offer a new *in vitro* approach to study *N. fowleri* infection in the context of primary amoebic meningo-encephalitis (Gianinazzi et al., 2005).

4.4. Ultrastructural features: amoebae from brain tissue *versus* culture medium

Using transmission electron microscopy, the changes of the ultrastructure of *N. fowleri* trophozoite in brain tissue of mice and culture medium revealed that (i) amoebae in mouse brain tissue were round in outline, whereas amoebae from axenic culture showed irregular appearance, (ii) mitochondria in the amoebae from axenic culture were oval, round and cylindrical shape and darkly stained, whereas amoebae from mouse brain tissue showed dumbbell shape together with above forms, (iii) rough endoplasmic reticulum of amoebae in brain tissue was tubular, but from culture it was vesicular or tubular in shape, (iv) empty vacuoles were demonstrated in amoebae from culture, while food vacuoles with myelinated structures were abundant in those from tissue, suggesting a strong phagocytic activity, (v) mouse brain tissue infected were extensively destroyed, and polymorphonuclear leukocytes were infiltrated predominantly with inflammatory lesion. Amoebae were observed in the vicinity of the capillary (Ryu et al., 1984).

4.5. Light and electron microsopic observations on the pathogenesis of *N. fowleri* in mouse brain and tissue culture

In mice (moribund mice) infected with *N. fowleri*, the normal architecture of the infected brain was completely destroyed; the olfactory lobes and the cerebral cortex showed the heaviest damage (Visvesvara and Callaway, 1974). The inflammatory response was mainly in the form of neutrophils and was confined to the olfactory lobes and the superficial regions of cerebral cortex. Numerous amoebae were seen interspersed with the degenerating neurones, glial processes, and neutrophils. Most conspicuous were the food vacuoles, which contained host tissue in various stages of digestion. Amoebae in the brain tissue also produced many micropinocytotic vesicles from the surface of the plasma membrane. These vesicles are interpreted as vehicles of transport of nutritive materials from the host tissue. Ultrastructural studies of the olfactory lobes from brains of dead animals revealed major concentrations of amoebae in the perivascular regions; amoebae were under attack by host polymorphonuclear leukocytes, and in the lumina of blood vessels (Schuster and Dunnebacke, 1977). Amoebae in brain tissue contained 30 nm intranuclear particles arranged in clusters. In the brains of some mice, particles and crystalloids were observed in the nuclei of degenerating cells of the central nervous system. *In vitro*, when Vero cell lines are infected with *N. fowleri*, host cells show cell shrinkage, nuclear pycnosis and discontinuity of cell sheet. Amoebae were often seen in an intracellular location. The Vero cells produced many fuzzy pinocytotic vesicles at these loci where the amoeba plasma membrane and Vero cell membrane were in close apposition. The pseudopodial formation and capturing of the host material indicated phagocytic activity of the amoebae. This was confirmed further by the presence of large numbers of food vacuoles containing host material in various stages of digestion suggesting that amoebae invade and destroy the brain tissue by active phagocytosis (Visvesvara and Callaway, 1974).

4.6. Routes of entry into the central nervous system

Experimentally, several routes have been tested including, intranasally, intraorally, into the conjunctival sac near the inner canthus of the eyes, and into induced skin lesions in adult germfree guinea pigs (Phillips, 1974). Of 33 animals inoculated intranasally with 18 to 31 amoebae, 31 developed a fatal encephalitis. There was considerable destruction of tissues of the cerebellum and the cerebrum and including the olfactory lobes. The meninges were involved to varying degrees in most of the animals. None of the animals inoculated by the three other routes developed either symptoms or lesions and it is likely that amoebae enter veins of the central nervous system and bone marrow during later stages of the disease (Phillips, 1974; Jaroli et al., 2002).

At 8 h post-infection, (i) amoebae are seen in contact with the mucous layer of the olfactory epithelium, (ii) numerous parasites eliminated by extensive shedding of the mucous layer, and (iii) many organisms reaching the nasal epithelium (Jarolim et al., 2000). At 24 h post-infection, infected mice exhibit focal inflammation and amoebae were observed in the submucosal nerve plexus, olfactory nerves penetrating the cribriform plate, and the olfactory bulb of the brain. The inflammatory response detected is scarce until 30 h post-inoculation. After 96 h, the inflammatory response was severe in the olfactory bulb and the brain, with tissue damage (Rojas-Hernández et al., 2004b).

4.7. Contact-dependent mechansims

For simplicity, the pathogenicity of *N. fowleri* is divided into contact-mediated and contact-independent mechanisms (Figure 4.1).

4.7.1. Adhesion

Adhesion is a primary step in amoebae-mediated host cell damage. The ability of trophozoites to attach to the nasal

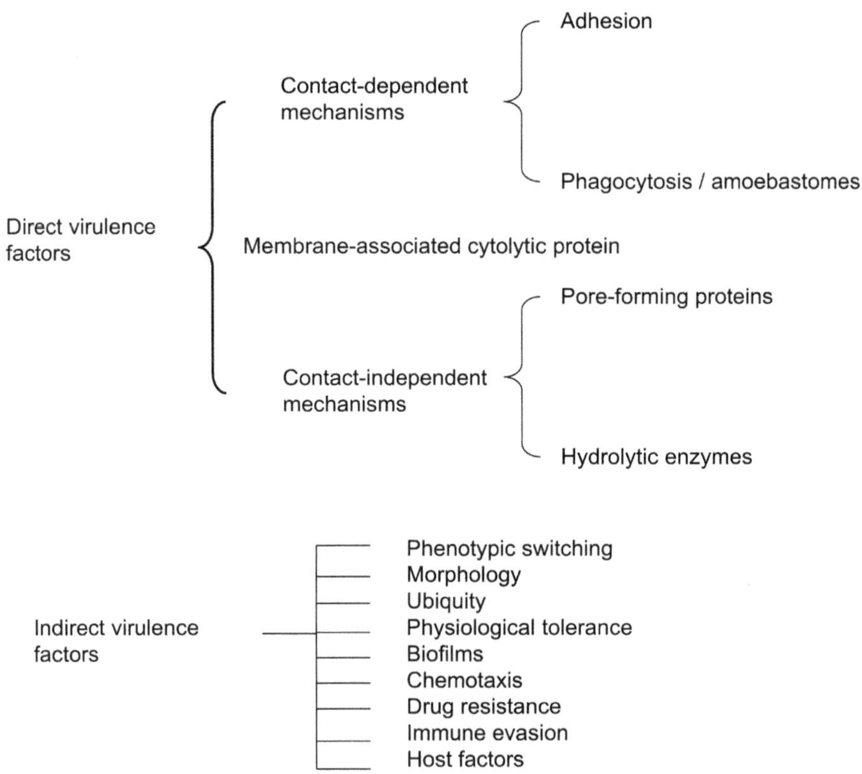

Figure 4.1. Factors associated with the pathogenesis of *N. fowleri* infection.

mucosa, an increased rate of locomotion, and a chemotactic response to nerve cell components are important in disease progression (Cline et al., 1986; Brinkley and Marciano-Cabral, 1992; Han et al., 2004). Like other pathogens, adhesion is mediated by adhesins expressed on the surface of the parasites. The possible role of glycoconjugates containing α-D-mannose and α-D-glucose residues in adherence of trophozoites to mouse nasal epithelium has been described by pre-treatment of trophozoites with three different lectins, and then inoculated intranasally in Balb/c mice. Mouse survival was 40% with lectins from pea (*Pisum sativum*) and Jackbean (*Canavalia ensiformis*) and 20% with lectins from snowdrop (*Galanthus nivalis*) amoebic pre-treatment, compared with 0% survival for control animals administered trophozoites without pre-treatment. These findings suggest that some of the glycoproteins found in *N. fowleri* represent an adherence factor (Carrasco-Yepez et al., 2013). *N. fowleri* exhibited a higher level of adhesion to the extracellular matrix components, laminin-1, fibronectin and collagen I. Scanning electron microscopy revealed that *N. fowleri* attached on extracellular matrix components substrata exhibited a spread-out appearance that included the presence of focal adhesion-like structures. Two integrin-like proteins are detected in *N. fowleri*. Confocal microscopy indicated that the integrin-like proteins co-localized to the focal adhesion-like structures. Furthermore, anti-integrin antibody decreased adhesion of *N. fowleri* to extracellular matrix components components (Jamerson et al., 2012).

A fibronectin binding protein of 60 kDa is found in extracts of *N. fowleri*. Western blot and immunolocalization assays using integrin alpha(5)/FnR antibodies showed that a 60 kDa protein reacted with the antibodies in extracts of *N. fowleri*, which was localized on the surface of *N. fowleri*. Pre-incubation of *N. fowleri* with the integrin antibodies significantly inhibited amoebic binding to fibronectin and cytotoxicity to the CHO cells. Additionally, protein kinase C activity was detected in the extract of *N. fowleri*. When *N. fowleri* was pre-treated with protein kinase C activator or inhibitor, the abilities of amoebic adhesion to fibronectin and cytotoxicity to the host cells were markedly affected compared to untreated amoebae. These results suggest that an amoebic integrin-like receptor and protein kinase C play important roles in amoebic cellular processes in response to fibronectin (Han et al., 2004). *N. fowleri* is also shown to induce reactive oxygen species (ROS) in host cells resulting in host cell death (Song et al., 2011).

4.7.2. Phagocytosis and amoebastomes

Both phagocytosis and amoebastomes are involved in amoebae-mediated host cell damage. When incubated with neuroblastoma cells, amoebae exhibited cytopathology from piecemeal consumption of target cells mediated by a sucker apparatus extending from the surface of *N. fowleri* (Marciano-Cabral and John, 1983). In co-culture, the target monolayer cells were damaged by phagocytosis by vigorous pseudopods and engulfment by sucker-like apparatus, amoebastomes (John et al., 1984; 1984b; Tiewcharoen et al., 2008). Both processes are actin-dependent that involve polymerization of monomeric G-actin into filamentous F-actin. A Nf-actin gene has been described to play a role in *N. fowleri* pathogenesis. The Nf-actin is localized on the cytoplasm, pseudopodia, and especially, food-cup structure (amoebastome) in *N. fowleri* trophozoites. Cytochalasin D, an actin polymerization inhibitor, blocked host cell cytotoxicity (Sohn et al., 2010).

Other studies identified a *nfa1* from *N. fowleri* that is 360 bp and encodes the Nfa1 protein (13.1 kDa) (Shin et al., 2001). The Nfa1 protein is located on the pseudopodia of *N. fowleri* trophozoites. The Nfa1 protein in *N. fowleri* trophozoites co-cultured with target cells is also located on pseudopodia, as well as in a food cup formed as a phagocytic structure in close contact with target cells. The amount of *nfa1* mRNA of

N. fowleri is strongly increased 6 h after co-culture (Kang et al., 2005). Notably, the anti-Nfa1 antibody reduces *N. fowleri*-induced host cell cytotoxicity (Cho et al., 2003; Jeong et al., 2004; Lee et al., 2007). Gene silencing of *nfa1* affects *N. fowleri*-induced host cell cytotoxicity (Jung et al., 2009). Anti-Nfa1 antibody reduced *N. fowleri*-mediated host cell pro-inflammatory response (Oh et al., 2005). Anti-Nfa1 antibody cannot detect Nfa1 protein expression in the *N. gruberi*, which also possesses *nfa1*. When cloned from *N. fowleri* and transfected into *N. gruberi*, it enhanced *N. gruberi*-mediated host cell cytotoxicity compared with naïve *N. gruberi* (Jeong et al., 2005; Song et al., 2006).

4.8. Membrane-associated cytolytic protein

Surface membrane-enriched fractions of *N. fowleri* contain a potent cytolytic activity. This surface membrane protein with cytolytic activity was unaffected by a treatment at 75°C for 30 min and accounted for 70 to 90% of cytolysis by whole-cell lysates of amoebae. The heat resistance as well as intimate membrane association distinguished the surface membrane cytolytic activity from a second heat-labile cytolytic activity which appears to be latent within lysosomes. The surface membrane cytolysin was found to be specifically activated by diluted samples of lysosomal fractions and may play a role in the pathogenicity of *N. fowleri* (Lowrey and McLaughlin 1985b).

A membrane protein, Mp2CL5 is expressed in pathogenic *N. fowleri* but not expressed in non-pathogenic *Naegleria* species nor in *Acanthamoeba*. The Mp2CL5 is a 17 kDa protein expressed on the plasma membrane of *N. fowleri* trophozoites. It is expressed in the logarithmic phase of trophozoite growth and the production of this protein increased through the stationary phase of growth (Réveiller et al., 2001; Zysset-Burri et al., 2014).

4.9. Contact-independent mechanisms

N. fowleri employs contact-independent mechanisms to produce host cell damage. The human microglial cells, co-cultured with *N. fowleri* trophozoites for 30 min in a non-contact system showed morphological changes such as the cell membrane destruction and a reduction in the number. A significant number of cells undergo death. These results suggest that the molecules released from *N. fowleri* in a contact-independent manner as well as phagocytosis in a contact-dependent manner may induce the host cell death (Kim et al., 2008c).

4.9.1. Pore-forming polypeptides

Earlier studies isolated a membrane-bound cytolytic pore-forming protein produced by *N. fowleri* called N-PFP. N-PFP was solubilized from amoeba membranes and exhibited a molecular mass of 66 kDa and 50-54 kDa, under reducing and non-reducing conditions respectively (Young and Lowrey, 1989). In addition to lysing erythrocytes, pore-forming protein is cytotoxic to several tumor cell lines. Its haemolytic activity is not dependent on the presence of divalent cations. The pore-forming protein depolarizes the membrane potential, suggesting that functional channel formation may represent the mode of membrane damage. The pore-forming protein is immunologically distinct from the pore-forming protein, perforin produced by lymphocytes, the terminal components of complement and pore-forming protein from *Entamoeba histolytica* (Young and Lowrey, 1989).

Later studies isolated two pore-forming polypeptides from extracts of *N. fowleri*. *N*-terminal sequencing and subsequent molecular cloning yielded the complete primary structures and revealed that the two polypeptides are isoforms. Both polypeptides share similar structural properties with antimicrobial and cytolytic polypeptides of *Entamoeba histolytica* (amoebapores) and of cytotoxic natural killer and T cells of human (granulysin) and pig (natural killer-lysin), all characterized by a structure of amphipathic alpha-helices and an invariant framework of cysteine residues involved in disulfide bonds. *Naegleria*pores A and B are processed from separate multipeptide precursor structures. According to their transcripts, each precursor molecule appears to contain additional *Naegleria* pore-like polypeptides, all of which share a structural motif of six invariant cysteine residues within their amino acid sequence. Moreover, biochemical characterization of the isolated polypeptides in combination with mass determination showed that they are *N*-glycosylated and variably processed at the *C*-terminus. The biological activity of the purified polypeptides of *Naegleria* was examined toward human cells and bacteria, and it was found that these factors, named naegleriapores, are active against both types of target cells (Herbst et al., 2002; Leippe and Herbst, 2004; Herbst et al., 2004).

4.9.2. Cytolytic activity of *N. fowleri* cell-free extract

The cell-free extract, prepared by lysing amoebae, was tested for cytotoxic activity against rat nerve cells. It elicited blebs on the target cell surface within 5 min after exposure to the fraction. Later, holes were observed in the plasma membrane. Phospholipase A, phospholipase C, sphingomyelinase, neuroaminidase, elastase, and other proteolytic enzyme activities are associated with the cell-free lysate. The cytotoxic activity was impaired by ethylenediaminetetraacetate (EDTA), phospholipase A inhibitor (Rosenthal's reagent), heating at 50°C for 15 min, or incubation at pH 10 for 60 min. Repeated freeze-thawing and inhibitors of proteolytic enzymes had no effect on the cytotoxic activity. Small amounts of ethanol (5% v/v) enhanced cytotoxic activity of the fraction (Fulford and Marciano-Cabral, 1986). In addition, *N. fowleri* express phospholipase A_2. Unlike the $cPLA_2$ enzyme in macrophages, this activity is largely calcium-independent, is constitutively associated with membranes and shows only a modest preference for phospholipids that contain arachidonate. *N. fowleri* PLA_2

activity is sensitive to inhibitors that block the activities of $cPLA_2$-alpha and the 80 kDa calcium-independent PLA_2, $iPLA_2$, that are expressed by mammalian cells (Barbour and Marciano-Cabral, 2001; Ferrante and Bates, 1988). In the brain section of primary amoebic meningoencephalitis patients, extensive demyelination is found in the white matter, besides the severe histopathological changes and large clusters of trophozoites in the gray matter. The myelinoclasis appears to be a result of a specific phospho-lipolytic effect, unlike that in post-viral encephalomyelitis, which has been attributed to vascular blockade or hemorrhages (Chang, 1979). Rat brain slices inoculated with *Naegleria* culture exhibited amoebic growth and demyelination. *Naegleria* degraded sphingomyeline with liberation of choline, sphingosine and fatty acids. These findings suggested that *N. fowleri* releases large amounts of phospholipases, lysophospholipase and sphingomyelinase, or factors, which "makes holes" in the lipid-rich cytoplasmic membrane of cells as well as causing demyelination of nerve tissue (Hysmith and Franson, 1982; Ferrante and Bates 1988).

4.9.3. Hydrolases

Hydrolases including proteases have been suggested to be involved in tissue invasion and destruction during infection and have been described in *N. fowleri* (Serrano-Luna et al., 2007). *N. fowleri* degraded zonula occludens-1 (ZO-1) and claudin-1 proteins but not occludin, likely mediated by extracellular proteolytic activity. In contrast, *N. gruberi* did not produce any damage (Shibayama et al., 2013). Using SDS-PAGE, protease profiles of total crude extract and conditioned medium of pathogenic *N. fowleri* and non-pathogenic *N. gruberi* showed differences in the number and molecular weight of proteolytic bands between the two strains. The proteases showed optimal activity at pH 7.0 and 35°C for both strains. Inhibition assays showed that the main proteolytic activity in both strains is due to cysteine proteases although serine proteases were also detected (Serrano-Luna et al., 2007). The presence of acid proteinase, *N*-acetylglucosaminidase, acid phosphatase, 5'-nucleotidase, aspartate aminotransferase, alpha-D-glucosidase, and an aminopeptidase have been demonstrated (Lowrey and McLaughlin, 1985). Among these, acid proteinase, *N*-acetylglucosaminidase, and acid phosphatase are associated with cytoplasmic granules closely resembling lysosomes; 5'-nucleotidase is associated with the surface membrane, probably on the external surface; aspartate aminotransferase is associated with mitochondria; and alpha-D-glucosidase and an aminopeptidase are associated with both the surface membrane and lysosomal particles (Lowrey and McLaughlin, 1985). In addition, cysteine proteases, lipases, sphingo-myelinase, elastase, cathepsin B-like proteases, beta-glucosidase, beta-galactosidase, beta-fucosidase, alpha-mannosidase, hexosaminidase, arylsulfatase A, and beta-glucuronidase, sphingomyelinase, neuraminidase, or arylsulfatase B, phospholipases, lysophospholipases, sphingomyelinases, neuraminidase, electrondense granules (small cytoplasmic components endowed with proteolytic activities), peroxiredoxin, thrombin receptor have been described, which may play a role in the pathogenicity of *N. fowleri* (De Jonckheere and Dierickx, 1982; Eisen and Franson, 1987; Ferrante and Bates 1988; Olomu et al., 1986; Kim et al., 2009; Chávez-Munguía et al., 2014).

Analysis of conditioned media of *N. fowleri* cultures revealed a major 30 kDa protease with substrate and inhibitor specificity consistent with cysteine proteases. Amino-terminal amino acid sequence of the purified enzyme showed it to be a thiol protease with homology to cathepsin L. It degrades extracellular matrix and produces cytopathic effect on mammalian cells. Both amoeba-induced matrix degradation and the cytopathic effect are inhibited by Z-Phe-Ala-fluoromethyl ketone, an irreversible cysteine protease inhibitor. Cysteine proteases of approximate molecular weights of 58 kDa, 128 kDa, and 170 kDa are described in *N.*

fowleri (Mat Amin, 2004; Vyas et al., 2014). Two cathepsin B and cathepsin B-like cysteine protease genes, cathepsin B (nfcpb) and cathepsin B-like (nfcpb-L) have been identified from cDNA library of *N. fowleri.* The full-length sequences of genes were 1,038 and 939 bp (encoded 345 and 313 amino acids), and molecular weights were 38.4 and 34 kDa, respectively. The rNfCPB and rNfCPB-L showed proteolytic activity for several proteins such as IgA, IgG, IgM, collagen, fibronectin, haemoglobin, and albumin. These findings showed that *N. fowleri* secretes a cysteine protease with the capacity to destroy host tissue (Aldape et al., 1994). These proteases play a role to acquire nutrients from different sources, including those from the host as well as producing damage to the human central nervous system (Serrano-Luna et al., 2007; Lee et al., 2014).

An intracellular alpha-aminoacyl-peptide hydrolase has been characterized. The enzyme preparation hydrolyzed phenylalanyl-, tyrosyl-, leucyl-, arginyl-, alanyl-, tryptophanyl-, histidyl-, methionyl-, and lysyl-naphthylamide but not benzoylleucyl-, leucylglycyl-, glycylprolylleucyl-, glycyl-, threonyl-, aspartyl-, or glutamyl-naphthylamide. The aminopeptidase activity was inhibited by the cysteine-protease inhibitors and by the metalloprotease inhibitor but not affected by the chelator, ethylenediaminetetraacetate, or the serine-protease inhibitor. The pH optimum was between 7.0 and 8.0. Enzyme activity was stable at 55°C (Marciano-Cabral et al., 1987b).

4.9.4. Nitric oxide

Although nitric oxide has a cytotoxic effect on pathogens, it can be produced as part of the pathology, as is the case with *Entamoeba histolytica*. The *N. fowleri* trophozoites produced nitric oxide *in vitro*. Western blot results showed *N. fowleri* trophozoites, contained proteins that share epitopes with the mammalian nitric oxide synthase, but have relative molecular weights different, suggesting that *N. fowleri* may contain undescribed NOS isoforms. Moreover, trophozoites reacted to the NOS2 antibody, in amoebic cultures as well as in the mouse brain infected with *N. fowleri*, suggesting that nitric oxide may participate in the pathogenesis of primary amoebic meningoencephalitis. (Rojas-Hernández et al., 2007).

4.9.5. Haemolytic activity

A haemolytic activity associated with postnuclear supernatant fractions of *N. fowleri* has been described. Haemolysis by *Naegleria* postnuclear supernatant fractions was sensitive to heat and trypsin hydrolysis, and was inhibited by divalent cations. The majority of the haemolytic activity was nonlatent and associated with a particle fraction sedimenting at 48,000 x g for 1 h. This particle-associated haemolytic activity appears to be membrane associated, as high salt concentration, chelating agents, and pH extremes were ineffective in solubilizing the haemolytic activity, whereas treatment with 0.15% Zwittergent 3-12, a dipolar ionic detergent, results in 98% release of the sedimentable haemolysin. The sigmoidal nature of the progress curve of postnuclear supernatant haemolysis, as well as synergistic interactions between fractions of amoebal whole cell extracts, suggests that the haemolytic activity has a multicomponent nature, with at least two and possibly three components participating in the haemolytic event (Lowrey and McLaughlin, 1984).

4.10. Additional potential pathogenicity factor

Heat shock protein 70 (HSP70) of *N. fowleri* has been cloned, and named as Nf-cHSP70. The Nf-cHSP70 is localized in the cytoplasm, pseudopodia, and phagocytic food-cups. The inhibition of synthesis of Nf-cHSP70 reduced the proliferation of *N. fowleri*. Nf-cHSP70 knock-downed *N. fowleri* with antisense oligomers showed reduced proliferation in comparison with untreated control, and reduced host cell cytotoxicity. These results suggest that Nf-cHSP70 plays an important role in the proliferation and may also play a role in the regulation of the host's immune system (Song et al., 2008; Zysset-Burri et al., 2014).

Moreover, cyclophilin is shown to strongly overexpressed in highly pathogenic *N. fowleri*, suggesting that this protein is an additional potential pathogenicity factor. Conversely, another protein, the HSP20 domain-containing protein, was over-expressed in weakly pathogenic *N. fowleri* (Zysset-Burri et al., 2014).

Apoptosis-linked gene-2-interacting protein X1 (AIP1) has been identified as a potential *N. fowleri* pathogenicity factor (Zysset-Burri et al., 2014). It is a key regulator of endosomal sorting, which accomplishes the intracellular transport of cellular material between organelles such as the Golgi apparatus as well as from organelles to the membrane and *vice versa* via vesicles. Yu et al., (2011) suggested that the Golgi-localized transmembrane protein HID-1, which is up-regulated in highly pathogenic *N. fowleri*, may be involved in vesicular exocytosis by preventing the mis-sorting of peptides to lysosomes for degradation (Yu et al., 2011; Wang et al., 2011). Zysset-Burri et al., (2014) suggested that both AIP1 and HID-1 are interesting candidate *N. fowleri* pathogenicity factors, potentially acting to regulate vesicular trafficking in the amoeba. Ras-related protein Rab-1, which was up-regulated in highly pathogenic compared to weakly pathogenic *N. fowleri*, may be involved in vesicular trafficking and, thus, in the phagocytosis of target cells (Zysset-Burri et al., 2014). Myosin (likely involved in phagocytic processes) is up-regulated in pathogenic *N. fowleri*. Villin-1 protein, likely involved in actin-dependent pathogenic processes, is found to be up-regulated in highly pathogenic *N. fowleri* compared to weakly pathogenic As villin-1 is involved in regulating the actin cytoskeleton and the fact that it showed the highest level of up-regulation in highly pathogenic *N. fowleri*, villin-1 is suggested as a promising candidate for further investigations to elucidate the molecular mechanisms involved in the pathogenesis of primary amoebic meningoencephalitis (Zysset-Burri et al., 2014).

5

The Host-Damage Response to *N. fowleri*

Abstract

Primary amoebic meningoencephalitis is a product of *N. fowleri* virulence factors and collateral damage from the host immune responses. Immune-mediated host damage is particularly important within the central nervous system, where the immune responses may exacerbate cerebral oedema and neurological damage, leading to coma and death. Given the challenges associated with the availability of effective antimicrobial chemotherapy, here we discuss existing knowledge of the role of immune response to *N. fowleri*. Consideration of the underlying mechanisms of host responses can provide critical insights into host damage that can be exploited to develop adjunctive therapies to improve disease outcome.

5.1. Role of immune response

Both humoral and cellular immunity contribute as the defence mechanism of the host against *N. fowleri*. Intravenous and intranasal immunization of mice with *N. gruberi* afforded 65% and 88% protection against subsequent intranasal challenge with *N. fowleri* suggesting the role of immune response in *N. fowleri* infection. Interestingly, non-pathogenic *N. gruberi* proved better immunogen than *N. fowleri*. Protection against *N. fowleri* can also be transferred in mice by immune serum (reviewed in John, 1982).

During the initial stages of infection, the host response is initiated by the secretion of mucus that traps the trophozoites on the surface of the nasal epithelia. Despite this response, some trophozoites are able to reach, adhere to and penetrate the epithelium. Mucins inhibited the adhesion of amoebae to host cells and reduced the cytotoxicity to target cells and the progression of the illness in mice. These studies suggested that mucus, including its major mucin component, may act as an effective protective barrier that prevents *N. fowleri* infection (Cervantes-Sandoval et al., 2008).

Once *N. fowleri* invades the tissue, it induces an inflammatory reaction (Rojas-Hernández et al., 2004b) (Figure 5.1). The contribution of the inflammatory response to brain damage in experimental primary amoebic meningo-encephalitis has been described. Several *N. fowleri* trophozoites were observed in the olfactory bulbs 72 h post-inoculation, and the number of amoebae increased rapidly over the next 24 h. Eosinophils and neutrophils surrounding the amoebae were noted at later times during infection. Electron microscopic examination of the increased numbers of neutrophils and the interactions with trophozoites indicated an active attempt to eliminate the amoebae. The extent of inflammation increased over time, with a predominant neutrophil response indicating important signs of damage and necrosis of the parenchyma. These data suggest a

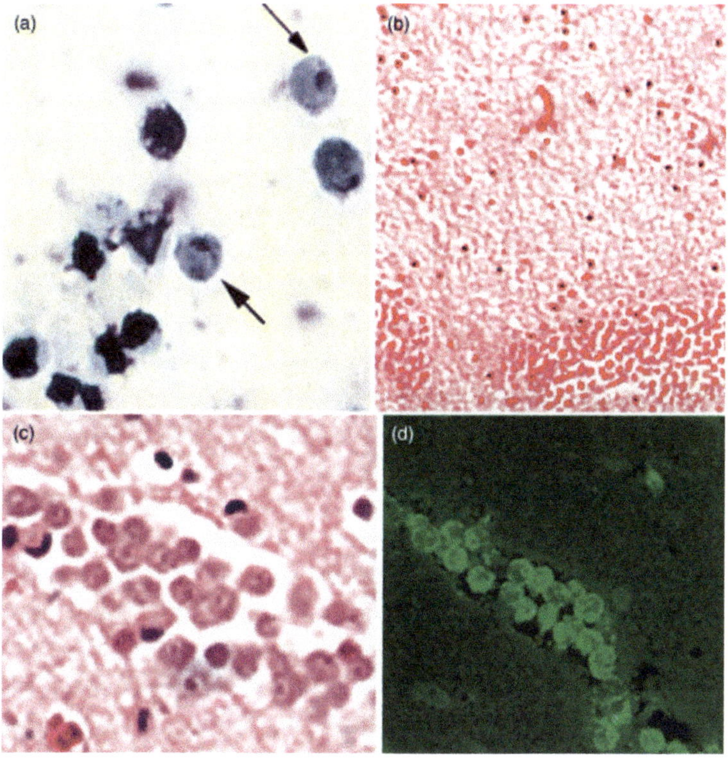

Figure 5.1. Pathological features of primary amoebic meningoencephalitis caused by *N. fowleri*. (a) A cerebrospinal fluid smear stained with Giemsa stain. Note amoebae at arrows (× 1000). (b) A section through the cerebellum showing extensive inflammation. (c) An area of the cerebellum showing extensive destruction of the brain architecture with large numbers of amoebae in the perivascular area (× 1000). (d) A section, similar to the one shown in (c), but reacted with anti-*N. fowleri* serum. Note the brightly staining *N. fowleri* trophozoites (kindly provided by G. S. Visvesvara from FEMS Immunol Med Microbiol. 2007, 50, 1-26).

probable role of inflammation in tissue damage. To test the former hypothesis, CD38-/- knockout mice with deficiencies in chemotaxis were used to compare the rate of mortality with the parental strain, C57BL/6J. The results showed that inflammation and mortality were delayed in the knockout mice. Based on these results, it was suggested that the host inflammatory response and polymorphonuclear cell lysis also contribute to the central nervous system tissue damage (Cervantes-Sandoval et al., 2008).

Using the human mucoepidermal cell line NCI-H292, it is shown that *N. fowleri* induced the expression of the MUC5AC gene and protein and the pro-inflammatory mediators interleukin-8, and interleukin-1 beta, but not tumour necrosis factor-alpha or chemokine c-c motif ligand 11 (eotaxin). The production of ROS is a common phenomenon involved in the signaling pathways of these molecules. Consistently, NCI-H292 cells generated reactive oxygen species within minutes of trophozoite stimulation. In addition, the use of an epidermal growth factor receptor inhibitor decreased the expression of MUC5AC and interleukin-8, but not interleukin-1 beta. It was concluded that *N. fowleri* induces the expression of some host innate defence mechanisms, such as mucin secretion (MUC5AC) and local inflammation (interleukin-8 and interleukin-1 beta) in respiratory epithelial cells via reactive oxygen species production (Cervantes-Sandoval et al., 2009).

Amoebae lysate incubation with rat microglial cells leads to pro-inflammatory cytokine release, including tumour necrosis factor-alpha, interleukin-6, and interleukin-1β. The generation of the former 2 cytokines was reduced with time, and that of the last increased throughout the experimental period. Overall, *N. fowleri* lysate exerted strong cytopathic effects on microglial cells, and elicited pro-inflammatory cytokine release as a primary immune response (Lee et al., 2011b).

Astrocytes are involved in the defence against infection and produce inflammatory responses. When incubated with human astroglial cells, *N. fowleri* lysates induced potent expression of interleukin-8, interleukin-1 beta, and interleukin-6. In addition, *N. fowleri* lysates induces the DNA binding activity of activator protein-1 (AP-1), an important transcription factor for interleukin-8 induction. Cytokine levels of astrocytes gradually decreased due to extracellular signal-regulated kinase (ERK), c-Jun N-terminal kinase (JNK) and p38 inhibitors. The transcription inhibitor (AP-1 inhibitor), downregulated interleukin-1 beta, interleukin-6, and interleukin-8 expression. These results show that AP-1 is related to interleukin-1 beta, interleukin-6, and interleukin-8 production. These findings show that *N. fowleri*-stimulated astrocytes lead to AP-1 activation and the subsequent expressions of interleukin-1 beta, interleukin-6, and interleukin-8, which are dependent on ERK, JNK and p38 MAPKs activation. These results imply that proinflammatory cytokines have important roles in inflammatory responses to *N. fowleri* infection (Kim et al., 2012; Kim et al., 2013).

5.2. Cell-mediated immunity

In an experimental model using ICR mice aging 6-7 weeks, delayed type hypersensitivity responses were observed following intranasal inoculation of parasites on day 1, 4, 7, 10 and 14 after infection. Concerning the blastogenic response of the splenocytes, it increased 10 days, post-infection of *N. fowleri*. The serum antibody titer of *N. fowleri*-infected mice increased from the day 7 and 14, post-infection respectively (Lee et al., 1989). In contrast, other studies using mice suggest that cell-mediated immunity does not play a role in protection against *Naegleria*. It has been shown that congenitally athymic mice (T cell deficient) and euthymic mice were equally susceptible to infection by *N. fowleri* (Newsome and Arnold, 1985). In addition, host resistance to *N. fowleri* infection was not affected when animals were treated with diethylstilbestrol, which depresses delayed typed hypersensitivity (Reilly et al., 1983b), suggesting that cell-mediated immunity alone does not constitute a major line of defense against *N. fowleri*. These findings led to conclusion that, as for humoral immunity, the role of cell-mediated immunity in protecting against infection by *N. fowleri* infection remains unclear (Marciano-Cabral and Cabral, 2007).

5.3. Neutrophils

Human neutrophils fail to kill the amoeba *in vitro*, but can do so if they are exposed to conditioned medium (CM) from Phytohaemagglutinin (PHA) stimulated mononuclear leucocytes. Specific antibody or complement was required to effect amoeba killing by CM modified neutrophils. Only short time exposure of the leucocytes to CM was required to endow them with amoebicidal properties (Ferrante and Mocatta, 1984). Likewise, neutrophils from *N. fowleri*-immunized mice are capable of killing amoebae. One method of killing is for a group of neutrophils to surround an amoeba and destroy it, presumably by contact and release of enzymes onto the amoeba cell membrane. However, a phagocytic process has been described in which neutrophils pinch off portions of an amoeba. Although unable to phagocytose an entire amoeba, several neutrophils are able to rupture an amoeba by pinching off and engulfing portions of it (reviewed in John, 1982; Ferrante et al., 1989). Tumour necrosis factor stimulated the adherence of neutrophils to *N. fowleri* with destruction of the amoeba. Notably, tumour necrosis factor-alpha augments the neutrophil activity by enhancing oxygen radical production, in response to *N. fowleri*. The priming effects of tumour necrosis factor-alpha were transient; marked enhancement was found with short 5- to 30-min pre-incubations of neutrophils with the cytokine. Tumour necrosis factor-alpha also augmented the neutrophil lysosomal enzyme release in response to *N. fowleri*. Neutrophils which lack myeloperoxidase but have a normal oxygen-dependent respiratory burst do not express the amoebicidal activity. The generation of amoebicidal activity required the presence of chloride ions. Arginine, a scavenger of hypochlorite, significantly depressed the ability of neutrophils to kill amoebae. Catalase inhibited the amoebicidal activity of neutrophils. These findings suggest a role for tumour necrosis factor-mediated destruction of *N.*

fowleri by neutrophils in host defense (Ferrante et al., 1987; Ferrante 1989; Michelson et al., 1990).

The role of neutrophils was further confirmed by treating immunized mice with the monoclonal antibody NIMP-R10, which binds to neutrophil complement receptor type 3bi (CR3) and causes selective neutrophil depletion in mice. Mice in the non-immunized group challenged with amoebae all died by day 12, while 97% in the immunized group survived. By contrast, the immunized group treated with NIMP-R10 showed only 25% survival. The immunized group treated with "control" mouse ascites, WEM-G11, was highly resistant (90% survival). There was a significant neutrophil response in the nasal mucosa and olfactory lobes of immunized, NIMP-R10-treated mice, despite a marked degree of neutropenia similar to that seen in immunized, untreated mice. Non-immunized mice showed virtually no neutrophil response. Despite this response in the NIMP-R10-treated mice, amoebic proliferation was not depressed, and there was no evidence of neutrophil degranulation or amoebic killing, despite the close apposition of large numbers of neutrophils to amoebae. The results indicate that neutrophils are necessary for the expression of immunity to *N. fowleri* (Ferrante et al., 1988).

5.4. Activated macrophages destruct *N. fowleri*

Activated macrophages mediate *N. fowleri* killing by producing nitric oxide in an arginine-dependent cytolytic mechanism and non-oxidative mediators including tumour necrosis factor-alpha and interleukin-1. However, the putative macrophage lytic molecule tumour necrosis factor-alpha is not amoebicidal because recombinant tumour necrosis factor-alpha alone, or in combination with interleukin-1 alpha or interleukin- beta, was neither cytolytic nor cytostatic for *N. fowleri*. Mouse peritoneal macrophages activated by different immunomodulators (*Mycobacterium bovis* or *Propionibacterium acnes*) destroyed *N. fowleri* amoebae by a contact-dependent process and by soluble cytolytic molecules secreted by macrophages in response to lipopolysaccharide. The inhibitors NG-monomethyl-L-arginine and arginase were used to determine whether the arginine pathway was a major effector mechanism responsible for amoebicidal activity of activated macrophages. Both the arginine analog NG-monomethyl-L-arginine and arginase, which breaks down arginine, decreased macrophage amoebicidal activity. Addition of arginine to arginine-free medium restores amoebicidal activity to activated macrophage cultures. These results demonstrate that the arginine pathway is an important mechanism for the destruction of susceptible *N. fowleri* amoebae (Fischer-Stenger et al., 1990; Fischer-Stenger and Marciano-Cabral, 1992). Cytolytic activity of the soluble amoebicidal factors can be recovered by ammonium sulfate precipitation of the conditioned media. Heat treatment of the conditioned media inactivated cytolytic activity suggesting that soluble proteins of activated macrophage are useful target for the amoebicidal activity (Cleary and Marciano-Cabral, 1986b).

5.5. T-lymphocytes

The amounts of T lymphocyte subsets and the blastogenic responses of T lymphocytes in mice after infected with *N. fowleri* showed that Thy 1.2+ T cells in the total spleen lymphocytes on day 7 were significantly increased compared with the control group. L3T4+ T cells and Ly2+ T cells in the total spleen lymphocytes on day 7 were significantly increased compared with the control group. The DNA S fraction of T cells in the spleen was significantly increased on day 7. The amount of S fractions of DNA was sequentially decreased on day 14 and 24, but they were also significantly increased compared with the control group. The interleukin-2 levels on day 14 of two experimental groups were significantly decreased as compared with the control group (Lyu et al., 1993).

5.6. Antibodies

Antibody to *N. fowleri* has been detected in surveys of normal human sera and saliva with titres ranging from 1:5 to 1:20 (Cursons et al., 1980b; reviewed in John, 1982). The antibodies belonged to IgA, IgG, and IgM. These antibodies could participate in the resistance to the infection, probably by inhibiting the adherence of *N. fowleri* trophozoites to the nasal mucosa (Cursons et al., 1980b; reviewed in John, 1982; Marciano-Cabral et al., 1987; Rivera-Aguilar et al., 2000; Rivera et al., 2001). This was further shown using *in vitro* studies, in which, sIgA antibodies were capable of inhibiting the adhesion of trophozoites to collagen type I (Shibayama et al., 2003). Later studies showed that antibodies inhibited proliferation of trophozoites, cytotoxicity of *N. fowleri* against host cells *in vitro* and *in vivo*, suggesting that antibodies weaken the virulence of the amoeba (Ryu and Im, 1991). *N. fowleri* trophozoites treated with antibodies showed an increased number of swollen mitochondria, disfigured cisternae, lipid droplets, and osmiophilic granules in the cytoplasm. The overall results suggest the protective effect of antibodies against *N. fowleri* infection (Soh et al., 1992).

5.7. Activation of complement

Neither *Naegleria* nor its culture supernatant was directly chemotactic for human neutrophils. However, interaction of *Naegleria* with human serum, resulted in the generation of a strong chemotactic stimulus (Rowan-Kelly et al., 1980). The reduction of serum activity by heat-inactivation indicated a dependence on serum complement for the interaction, suggesting the involvement of complement in *N. fowleri* infection. Later studies showed that *N. fowleri* were lysed by adult fresh human serum, and their multiplication was inhibited in culture medium supplemented with 10% fresh human serum (Holbrook et al., 1980). Heat inactivation (56°C, 30 min) of serum abrogated these lytic and inhibitory effects. Conversion of C3 and C3i occurred after incubation of *N. fowleri* with serum which had been treated with ethylene glycol-bis(beta-aminoethyl ether)-N,N-tetraacetic acid, indicating activation of complement via the alternative pathway (Holbrook et al., 1980). The presence of specific agglutinating antibody to *N. fowleri* enhanced the amoebicidal activity of serum for *N. fowleri* (Whiteman and Marciano-Cabral, 1987).

5.8. Natural killer cell

The natural killer cell activity was examined after inoculating *N. fowleri*. The natural killer cell activities increased significantly on day 1 after infection as compared with the control group, and then remarkably declined thereafter. Maximal killing potential, maximal recycling capacity and percentages of activated natural killer cells were remarkably increased in infected mice as compared with the control. There was a significant increase in the cytotoxic activity of natural killer cells in the infected group (Lee et al., 1991; Kim et al., 1993).

5.9. Immune evasion

Although mucus, including its major mucin component, can act as an effective protective barrier, *N. fowleri* re-establish the capacity to adhere. In addition, mucinolytic activity in *N. fowleri* has been demonstrated and a 37 kDa protein with mucinolytic activity has been identified. The activity of this protein can be inhibited by cysteine protease inhibitors. These findings suggest that when the number of amoebae is sufficient to overwhelm the innate immune response, the parasites may evade the mucus by degrading mucins via a proteolytic mechanism (Cervantes-Sandoval et al., 2008).

Following infection, there was an impairment of the blastogenic response of splenocytes to *N. fowleri* during the acute course of experimental primary amoebic meningo-encephalitis in mice (Park et al., 1987). Lymphoblastic transformation induced by T-cell mitogen was markedly reduced in comparison to the uninfected control mice. The blastogenic response to B-cell mitogen remained depressed in the infected mice up to 14 days after infection suggesting that there is a suppression of cell mediated immunity during the acute course of experimental *Naegleria* meningoencephalitis in mice (Im et al., 1987).

Weakly pathogenic and non-pathogenic *Naegleria* spp. are lysed by complement but highly pathogenic *N. fowleri* are resistant to complement-mediated lysis. *In vivo* studies showed that mice depleted of complement by cobra venom factor are more susceptible to infection (Reilly et al., 1983). Pathogenic *N. fowleri* maintained in Cline medium are virulent for mice and resistant to complement lysis. A rapid decline in resistance to complement and virulence for mice is observed when pathogenic *N. fowleri* are grown in Nelson medium lacking hemin. *N. fowleri* maintained in Nelson medium can be rendered complement-resistant by shifting the amoebae to growth in Cline medium. Cycloheximide treatment of *N. fowleri* maintained in Nelson medium blocks the transition to a complement-resistant phenotype (Toney and Marciano-Cabral, 1994b).

In addition, the effect of effect of Ca^{2+} ions in initiating complement resistance of *N. fowleri* has been studied. Chelation of extracellular calcium with ethylene glycol tetraacetic acid (EGTA) or chelation of intracellular calcium with 1,2-bis-(O-Aminophenoxy) ethane-N,N,N,N tetraacetic acid tetra (acetoxymethyl) ester (BAPTA-AM) increased complement lysis of *N. fowleri*. Increased lysis of ionomycin-treated *N. fowleri* was detected after exposure to serum complement, suggesting that a threshold level of Ca^{2+} mediates complement resistance before survival mechanisms are overwhelmed and lysis occurs. Overall, these data indicate that Ca^{2+} ions influence complement resistance in *N. fowleri* (Chu et al., 2002).

In vitro studies have shown that pathogenic *N. fowleri* employ at least two means to resist complement damage: (i) expression of complement-regulatory proteins and (ii) shedding of the membrane attack complex (C5b-C9) on vesicles (Toney and Marciano-Cabral, 1992, 1994). Treatment of trophozoites with papain or trypsin, but not with neuraminidase, increased susceptibility of highly pathogenic *N. fowleri* to complement lysis. Treatment of trophozoites with actinomycin D or cycloheximide during incubation with normal human serum or pre-treatment with various protease inhibitors for 4 h did not increase the susceptibility of *N. fowleri* amoebae to lysis. Neither a repair process involving *de novo* protein synthesis nor a complement-inactivating protease appears to account for the increased resistance of *N. fowleri* amoebae to complement-mediated lysis (Whiteman and Marciano-Cabral, 1989), except the likely expression of complement-regulatory proteins. In support, *N. fowleri* is shown to express a "CD59-like" surface protein (Fritzinger and Marciano-Cabral, 2004). In mammalian cells, CD59 is a complement-regulatory protein that inhibits complement-mediated lysis of cells, which may explain yet another mechanisms of complement evasion by *N. fowleri*. CD59-like protein (a 18 kDa) is expressed on the membrane of *N. fowleri*. Complement component C9 was immuno-precipitated *N. fowleri* CD59-like protein, when amoebae were incubated with normal human serum. Collectively, these studies suggest that a protein reactive with antibodies to human CD59 is present on the surface of *N. fowleri* amoebae and may play a role in resistance to lysis by cytolytic proteins (Fritzinger et al., 2006).

With regards to shedding of the attack complex, membrane blebbing was observed on the surface of complement-resistant *N. fowleri* in response to incubation in normal human serum diluted 1:4 to 1:16. The membrane attack complex was concentrated on the membrane blebs. Treatment of complement-resistant *N. fowleri* with cytochalasin D or cytochalasin B to inhibit actin-dependent exocytic processes increased the susceptibility of the amoebae to complement damage. These data suggested that pathogenic *N. fowleri* use membrane vesiculation to remove membrane-deposited complement proteins, specifically the membrane attack complex (C5b-C9). The ability to remove surface-associated membrane attack complexes is an important mechanism by which pathogenic *N. fowleri* resist complement lysis (Toney and Marciano-Cabral, 1994). Enzymatic removal of surface components from pathogenic *N. fowleri* with phosphatidylinositol-specific phospholipase C or with endoglycosidase H increased the susceptibility of amoebae to complement-mediated lysis. Tunicamycin treatment of pathogenic *N. fowleri* increased susceptibility to lysis by complement. Proteins of 234 and 47 kDa were detected in supernatant fluid from phosphatidyl-inositol-specific phospholipase C-treated pathogenic amoebae. The surface proteins of pathogenic *N. fowleri* are 89, 60, 44, and 28 kDa glycoproteins (Toney and Marciano-Cabral, 1992).

When exposed to serum, activation of protein kinases including serine/threonine or tyrosine kinases and subsequent protein phosphorylation is observed that are important in mediating complement resistance in *N. fowleri* (Chu et al., 2000). Inhibition of protein kinases makes them susceptible to complement lysis. Overall, these results suggest that kinases play a critical role in the protection of *N. fowleri* to complement-mediated lysis. The "CD59-like" protein is also expressed when *N. fowleri* is cocultured with bacteria that produce toxins (Fritzinger and Marciano-Cabral, 2004). Fritzinger et al., (2006) suggested that the "CD59-like" protein could protect amoebae, not only from the action of the membrane attack complex of complement (C5b-C9) but also from that of pore-forming proteins such as bacterial toxins and the pore-forming proteins produced by *N. fowleri* known as naegleriapores.

Macrophage inhibition by the lymphokine macrophage inhibition factor has been described for *N. fowleri*. *N. fowleri* internalize surface-bound antibody that may enable them to evade the host's immune defenses (reviewed in John, 1982).

Although sIgA antibodies capable of inhibiting amoebae adhesion, parasites are capable of eliminating the antigen-antibody complex produced on the surface (Shibayama et al., 2003). *N. fowleri* was observed to cap and internalize surface-bound antibody (Ferrante and Thong, 1979). These results suggest that the ability of *N. fowleri* to remove antibody from its surface may allow the amoeba to resist the action of the host's immune system.

5.10. Immunization using whole parasites

Intravenous, intranasal, subcutaneous, and intraperitoneal immunization of mice with formalin/formaldehyde-fixed *N. gruberi* afforded protection against subsequent intranasal challenge with *N. fowleri* (reviewed in John, 1982). In general, intravenous inoculation afforded greater protection than other routes of immunization. Intact parasites immunized mice better than did cell fragments, and *N. gruberi* appeared to be a better immunogen than *N. fowleri*. Other studies reported that mice immunized with live *N. fowleri* intraperitoneally were more resistant to subsequent intranasal challenge (John et al., 1977; Thong et al., 1978; Lee et al., 1985; Bush and John, 1988). Cry1Ac protoxin protein, produced by the bacterium *Bacillus thuringiensis*, is used as a mucosal adjuvant (an immune-response enhancer). It has been tested to develop a vaccine against *N. fowleri* (Rojas-Hernández et al., 2004). Notably, amoebae lysates co-administered with Cry1Ac increased protective immunity against fatal *N. fowleri* infection in mice to 100%. Cry1Ac protoxin has potent mucosal and systemic adjuvant effects on antibody responses to proteins or polysaccharides. When mice intranasally immunized with Cry1Ac plus lysates are challenged with amoebae, both IgG and IgA mucosal responses are rapidly increased, but only the increased IgG response persisted until day 60 in surviving mice (Rojas-Hernández et al., 2004). In addition. the immunization provoked an increase in areas with metaplasia in the olfactory epithelium, allowing for secretion of IgA that could lead to greater protection against *N. fowleri.* The increased IgA response induced in the nasal mucosa by immunization likely impedes both amoebic adhesion and subsequent invasion of the parasite to the nasal epithelium (Jarillo-Luna et al., 2007). Later studies showed that Cry1Ac induced significant specific IgA and IgG cell responses, especially in nasal passage. Besides, it increased the proportion of lymphocytes expressing the activation markers CD25 and CD69 in nasal passage and nasal-associated lymphoid tissue. CD25 was increased in B cells along with CD4 and CD8 T cells from nasal-associated lymphoid tissue and nasal passage, while CD69 was increased in B cells from both tissues but only in CD4 T cells from nasal passage. Cry1Ac augmented especially a Th2 profile of cytokines, as the proportion of T cells that spontaneously produced interleukin-4, interleukin-5 and interleukin-10 was increased and this effect was higher in nasal passage than in nasal-associated lymphoid tissue (Rodriguez-Monroy and Moreno-Fierros, 2010). Following immunization with amoebic lysates plus Cry1Ac, the resistance to *N. fowleri* infection is attributed to STAT6-induced Th2 immune response. When immunized intranasally with amoebic lysates plus Cry1Ac, STAT6-deficient (STAT6-/-) mice showed 0% protection against subsequent challenge with *N. fowleri.* In contrast, STAT6+/+ mice displayed 100% protection. Significantly higher titres of Th2-associated IgG1 as well as interleukin-4 were found in STAT6+/+ mice, whereas in STAT6-/- mice significantly more interleukin-12 and interferon-gamma as well as significantly higher titres of Th1-associated IgG2a were detected. Thus, whereas protected STAT6+/+-immunized mice elicited a Th-2 type immune response that produced predominantly humoral immunity, unprotected STAT6-/- mice exhibited a polarized Th1 type cellular response. These findings suggest that the STAT6-signalling pathway is critical for defence against *N. fowleri* infection (Carrasco-Yepez et al., 2010).

In addition to Cry1Ac, Cholera toxin is also a useful adjuvant. Immunization with *N. fowleri* lysates plus Cholera toxin increased the protection (survival up to 100%) against *N. fowleri* infection in mice and antibodies IgA and IgG together with polymorphonuclear cells avoid the attachment of *N. fowleri* to apical side of the nasal epithelium. The nasal immunization resulted in the induction of antigen-specific IgG subclasses (IgG1 and IgG2a) in nasal washes at days 3 and 9 after the challenge and IgA and IgG in the nasal cavity, compared to healthy and infected mice. Immunization increased the expression of the genes for alpha chain, its receptor (pIgR), and it also increased the expression of the corresponding proteins evidenced by the ~65 and ~74 kDa bands, respectively. Immunization resulted in an increased expression of interleukin-10, interleukin-6, and interferon-gamma cytokines. Immunization inhibited the production of tumour necrosis factor-alpha compared to the infected group where the infection without immunization caused an increase in it. The co-existence of selected cytokines produced by immunization may provide a highly effective immunological environment for the production of IgA, IgG and pIgR as well as a strong activation of the polymorphonuclear leukocytes in mucosal effector tissue such as nasal passages (Carrasco-Yepez et al., 2014).

5.10.1. Immunization using cell supernatants

Later studies show that immunization with multiple doses of *N. fowleri* culture supernatant produced a survival of 67 to 78%. Fractionation of the culture supernatant by column chromatography showed that all six fractions contained protective antigens, but the best protection occurred from immunization with the high molecular weight fraction (greater than 200 kDa). The degree of protection conferred to mice is related to the levels of anti-*Naegleria* antibodies. These antibodies react with the surface of the amoeba. The data also show that serum (and the IgG serum fraction) from immunized mice confer protection to normal mice against a lethal *N. fowleri* challenge. Spleen cells from immunized animals were only capable of conferring protection to recipients, when the challenge time was delayed (10 days), at which time anti-*Naegleria* antibodies appeared in the serum of the mice. Histological observations suggested that this protection is expressed at the nasal mucosa and possibly results from the combined effects of polymorphonuclear leucocyte-mediated killing of the amoeba and mechanical elimination of the organisms by extensive shedding of necrotic epithelium. Overall, these studies suggested that *N. fowleri* culture supernatant induced resistance to *N. fowleri* meningoencephalitis ((Thong et al., 1983; Ferrante and Rowan-Kelly 1988).

5.10.2. Passive immunity

As the majority of cases prove fatal, intracisternal passive immune therapy in rabbits with amoebic meningoencephalitis by using anti-*Naegleria* immune serum, an immunoglobulin G fraction, and a monoclonal antibody to *N. fowleri* were tested. The immune serum and the immunoglobulin G fraction isolated from it by affinity chromatography or the monoclonal antibody provided a consistent, although temporary, protective effect, shown by prolongation of survival. Multiple doses of immune serum further prolonged survival. The protective effect of serum was retained after heating to 56°C. These findings suggest that the passive intracisternal antibody therapy might serve as an adjunctive component in the treatment of primary amoebic meningoencephalitis (Lallinger et al., 1987).

5.10.3. Immunization with the rNfa1 protein

The rNfa1 protein is important in the pathogenicity of *N. fowleri* and it has been tested as a vaccine target. Mice immunized intra-peritoneally or intra-nasally with rNfa1 protein developed specific IgG, IgA and IgE antibodies; the IgG response was dominated by IgG1, followed by IgG2b, IgG2a and IgG3. High levels of the Th1 cytokine, interferon-gamma, and the regulatory cytokine, interleukin-10, were also induced. The mean survival time of mice immunized intra-peritoneally with rNfa1 protein was prolonged compared with controls, (25.0 and 15.5 days, respectively). Similarly, the mean survival time of mice immunized intra-nasally with rNfa1 protein was 24.7 days, compared with 15.0 days for controls (Lee et al., 2011).

Later studies tested nfa1 DNA vaccination in mice. To evaluate immune responses of nfa1-vaccinated mice, mice were intranasally vaccinated with viral particles of nfa1 gene. The levels of both IgG and IgG subclasses (IgG1 and IgG2a) in vaccinated mice were significantly increased. The cytokine analysis showed that vaccinated mice exhibited greater interleukin-4 and interferon-gamma production than the other control groups, suggesting a Th1/Th2 mixed-type immune response. In vaccinated mice, high levels of Nfa1-specific IgG antibodies continued until 12 weeks post-vaccination. The mice vaccinated with viral vector expressing the nfa1 gene also exhibited significantly higher survival rates (90%) after challenge with *N. fowleri* trophozoites. The nfa1 vaccination induced protective immunity by humoral and cellular immune responses in *N. fowleri*-infected mice. These results suggest that DNA vaccination using a viral vector may be a potential tool against *N. fowleri* infection (Kim et al., 2012b; Kim et al., 2013b).

6

Cell Biology and Speciation

Abstract

N. fowleri is a free-living opportunistic protist. Although the ability of *N. fowleri* to produce infection of the central nervous system has gained significant attention as an important human pathogen, producing fatal primary amoebic meningoencephalitis; however, it also has a fascinating biology exhibiting three life forms, and may play an important role in the ecosystem. The study of the cell and molecular biology of ancestral eukaryotic single-celled protists such as *Naegleria*, can provide fundamental insight of molecular pathways that evolved in multicellular organisms, as well as understanding of the causal relationships between genotype and phenotype.

6.1. Discovery of *N. fowleri*

N. fowleri was first discovered by Fowler and Carter in Australia (Fowler and Carter 1965) as the causative agent of primary amoebic meningoencephalitis. *N. fowleri* was named in honour of Dr Malcom Fowler who first recognized the disease. Among the genus *Naegleria*, this is the only pathogenic species to be isolated from primary amoebic meningoencephalitis patients.

6.2. Different life forms of *N. fowleri*

N. fowleri has three life forms, an amoeboid form, a flagellate form, and a cyst form (Figure 6.1). The flagellate form is a transient, non-feeding, non-dividing form. Notably, the flagellate form neither divides nor encysts. Only the amoeboid form is able to feed, divide, and encyst. Cyst form excysts into the amoeboid form. The three forms and their inter-conversion make this protist an excellent model to study cellular differentiation processes.

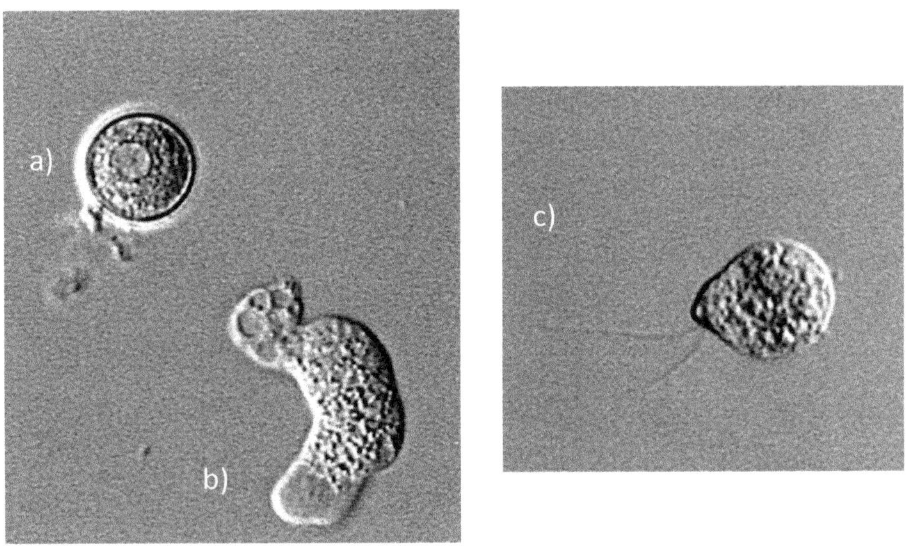

Figure 6.1. *N. fowleri* (a) cyst form, and (b) trophozoite form, and (c) flagellate form, viewed under phase contrast microscopy. X1000. (courtesy: G. S. Visvesvara, CDC, USA).

6.3. Ultrastructural analysis

The amoeboid form is characterized by a large nucleus with a nucleolus, that is typically vesicular, the presence of numerous mitochondria, food vacuoles, a contractile vacuole, endoplasmic reticulum enclosed within a plasma membrane of 10 nm in width (Fulton, 1970; Schuster, 1963). A large number of ribosomes (free and attached to the endoplasmic reticulum) are scattered throughout the cytoplasm. A smooth endoplasmic reticulum is also present. Numerous membrane-bound cytoplasmic organelles and dumbbell-shaped mitochondria are observed. The resting nucleus is spherical. A sharply defined nuclear division is promitotic, in which the nucleolus elongates and divides into two polar masses and the nuclear membrane remains intact. The amoeboid form is long and slender (8-15 μm) and move by forming one or more lobose pseudopodia (Schuster, 1963; Forrester et al., 1967).

Cysts are spherical, with a smooth single layer, often clumped closely together, and 7-12 μm in diameter. Ultrastructure examination reveals an average of one to two mucoid-plugged pores per cyst through which regenerated trophozoite emerges (John, 1982). The cyst wall of *N. gruberi* contain a high protein content, readily extractable lipid component, and an alkali-insoluble material that is suggested to be cellulose. The high lipid and protein content and masking of the cellulose-staining properties in the unhydrolyzed cyst wall suggest that the cellulose may be bound to a lipoprotein complex (Werth and Kahn, 1967).

The flagellate has an elongate, pear-shaped body, usually possessing two flagella of equal length, a nucleus in the narrower anterior region, and no cytostome. The ultrstructure of *N. fowleri* flagellates is that of a typical eukaryotic protist. There is a distinct nuclear membrane and prominent nucleolus, numerous vacuoles and cytoplasmic inclusions, pleomorphic mitochondria, and rough endoplasmic reticulum.

The kinetic apparatus consists of flagella and their associated kinetosomes, a "spur" made up of isodiametric fibrils (14 µm in diameter) attaching to the base of the kinetosome, a cross-striated rhizoplast, and thin-walled fibrils (22 µm in diameter) with a center of attachment in the kinetosomal complex. Notably, neither centrioles nor kinetosomes are found in the amoeba form. Complex arrangements of endoplasmic reticulum are seen in the cytoplasm of transforming *Naegleria* (Fulton, 1970; Schuster, 1963). The nuclei in metamorphosing *Naegleria* frequently show unusual plasticity (Schuster, 1963).

6.3.1. Centrin, centrioles and microtubule-organizing centers (MTOCs)

Centrin is an approximately 20 kDa calcium-binding protein. It is a component of centrosome-associated flagellar roots capable of calcium-mediated contraction, and is also found in the centrosomes of vertebrate cells. The analysis of a centrin gene from *Naegleria* revealed conserved features that distinguish centrins from calmodulin. In *Naegleria* flagellates, centrin is associated intimately with the cylinder of the basal bodies (Levy et al., 1996).

6.3.2. Nucleolar protein BN46/51

BN46/51 is an acidic protein found in the granular component of the nucleolus of *Naegleria*. When *Naegleria* amoebae differentiate into swimming flagellates, BN46/51 is found associated with the basal body complex at the base of the flagella. BN46/51 represents a unique nucleolar protein that can form specific complexes with fibrillarin and other nucleolar proteins. It was suggested that the association of BN46/51 with the MTOC of basal bodies may reflect its role in connecting the nucleolus with the MTOC activity for the mitotic spindle. This would provide a mechanism for nucleolar segregation during the closed mitosis of *Naegleria* amoebae (Trimbur et al., 1999).

6.3.3. Flagellar rootlet of *Naegleria*

Attached to the basal body is a striated rootlet or rhizoplast. Rootlets isolated from mature flagellates are approximately 13 µm long but vary from 8 - 15 µm in length (Larson and Dingle, 1981). Rootlets are composed of filaments, approximately 5 nm diameter, associated in a linear fashion to yield the characteristic 21-nm cross-banded appearance. Differential solubilization of rootlets and their associated contaminants allowed identification of a major rootlet protein, comprising at least 50% of any purified rootlet preparation, with an apparent subunit molecular weight of 170 kDa (Larson and Dingle, 1981). Gardiner et al., (1981) isolated the rhizoplast from the flagellate stage of *N. gruberi*. The rhizoplast is proteinaceous. Polyacrylamide gel electrophoresis showed subunit molecular weight of approximately 240 kDa. Studies with antisera raised against the rhizoplast fraction demonstrated the absence of rhizoplast antigens in amoeboid form suggesting that the organelle is synthesized *de novo* during transformation of the amoeba to the flagellate form (Gardiner et al., 1981).

6.3.4. Flagellar tubulin

The flagellar tubulin is similar to other tubulins in molecular weight (55 kDa), amino acid composition, electrophoretic mobility, and nucleotide composition (Kowit and Fulton, 1974). The measurements indicate that the tubulin increases 35- to 55-fold during the differentiation of amoebae to flagellates (Kowit and Fulton, 1974).

6.3.5. Microfilaments

Electron microscopy has revealed 2 types of microfilament in the cytoplasm of *N. fowleri* (Lastovica, 1976). Thin, actin-like microfilaments, 5-7 nm in diameter are randomly oriented in the nonmotile amoebae, and are concentrated near the plasma membrane. In the actively motile amoebae, these microfilaments aggregate to form co-lateral bundles in close proximity to the plasma membrane. Thick filaments (myosin-like filaments) 17-19 nm in diameter also occur in the amoebae cytoplasm (Lastovica, 1976). The purified actin shared many attributes with numerous other actins that have been characterized, including molecular weight, strong binding to DEAE-cellulose, binding to DNase I, reversible polymerization to F-actin, binding of rabbit myosin subfragment 1 to give distinctive arrowheads, formation of Mg paracrystals, and activation of myosin Mg^{2+}-ATPase. In two respects, the attributes of *Naegleria* actin are unusual. Isoelectric focusing resolves three distinct isoforms of the actin, which raises questions about the function of multiple isoforms in a unicellular eukaryote. The amino acid composition closely resembled other actins except that *Naegleria* actin lacks *N* tau-methylhistidine. It was suggested that *N* tau-methylhistidine is not a prerequisite for actin-actin or actin-myosin interactions (Sussman et al., 1984). Later studies cloned and sequenced an intronless actin gene from *N. fowleri*. Codon usage and third-position-codon nucleotide frequency were significantly different from *Acanthamoeba*. Between the two amoebae, actin peptide sequences were 92.8% similar, while nucleotide sequences were only 70% similar. A phylogenetic reconstruction of actin amino acid sequences, using a distance method, placed *Naegleria* in a cluster with *Plasmodium* and *Entamoeba* (Gorospe et al., 1996).

6.3.6. Actomyosin complex

A protein complex similar to muscle actomyosin and plasmodial myosin B has been isolated from genus *Naegleria* (Lastovica and Dingle, 1971). It comprises approximately 0.7% of the total cell protein, had the solubility properties of actomyosin, displayed Ca^{2+}-activated, Mg^{2+}-inhibited ATPase activity, and forms microfilaments. Both ATPase activity and superprecipitation of the actomyosin complex were inhibited by the sulfhydryl inhibitor salyrgan (Lastovica and Dingle, 1971).

6.4. Motility

N. fowleri exhibits rapid locomotion at 37°C (Thong and Ferrante, 1986). On an agar plate with *E. coli*, *N. fowleri* exhibits growth and movement at the rate of 1 to 3 mm per day at 23°C, 7 to 14 mm per day at 37°C, and 7 to 14 mm per day at 43°C (Rowbury et al., 1983). The major manifestations of amoeboid locomotion in *Naegleria*-cytoplasmic streaming, pseudopod production, cell polarity and focal contact production require dynamic actin-based cytoskeleton. *Naegleria fowleri* amoebae demonstrate a chemotactic and chemokinetic response towards live cells and extracts of *Escherichia coli* and other bacterial species. The peptide *N*-formyl-methionyl-leucyl-phenylalanine acted as a chemokinetic rather than a chemotactic factor for *N. fowleri* amoebae (Marciano-Cabral and Cline, 1987). Competition experiments in which nerve cell extracts or bacteria are placed on either side of the filter in chemotaxis chambers result in increased movement towards bacteria than nerve cells. A scanning electron microscopy study of the interaction of *N. fowleri* with different bacterial species confirmed that when the amoebae were near ingestible bacteria they moved toward the bacteria by pseudopod formation. *Naegleria fowleri* appeared to respond to bacteria by three interrelated but distinct processes: chemokinesis, chemotaxis, and formation of food cups (Marciano-Cabral and Cline, 1987).

For locomotion on a glass substrate, two types of contact are made with a planar glass substrate. One, formed at a considerable distance from the substrate in deionized water (congruent to 100 nm), termed as "associated contact". From this platform filopodia are produced, which form close contacts, termed as "focal contacts" (Preston and King, 1978). In locomotion the area of associated contact is very mobile, in contrast to the focal contacts which, once established, are stable. Focal contact sites are left behind on the glass surface ("footprints") when the amoeba moves away. The cell-substrate gap in the associated contact is greatly affected by the ionic strength of the medium and particularly the valence of the cation component. This suggests that long-range forces of attraction play an important role in keeping the amoeba close to a substrate and thus allow the production of filopodia from the ventral surface to form focal contacts (Preston and King, 1978).

Naegleria is able to carry out amoeboid locomotion at the water-air interface in a manner indistinguishable from that exhibited on solid substrata with the production of focal contacts and associated filopodia. The speed of locomotion at this interface can be modulated by changes in electrolyte concentrations; these speed changes are identical to those observed at a water-glass interface. These findings suggest that the surface tension alone could provide suitable properties for the adhesion and translocation of amoebae at this interface without necessitating specific, absorbed molecules. The temporary swimming flagellate stage of *Naegleria* is able to dock at the interface, make stable adhesions to it, and revert to the amoeboid phenotype. Conversely, amoebae resident at the water-air interface can transform to swimming flagellates and escape into the bulk liquid phase (Preston and King, 2003).

6.5. Biochemical composition

The biochemical cell composition of *N. fowleri* is related to culture age. For agitated axenic cultures, average dry cell mass is constant during log growth at 150 pg per amoeba, but decrease 30%, during stationary growth at 96 h. During log growth, 80-85% of the dry cell mass is protein (120 pg per amoeba). Dry cell mass and protein of *N. fowleri* are about 70% of values reported for *N. gruberi*. During log and stationary growth phases, carbohydrate content is approximately 15 pg per amoeba, and RNA is about 18 pg per amoeba. Total DNA content is 0.2 - 0.3 pg per amoeba during log growth, but it increases during transition from log phase to stationary phase and then gradually decreases to nearly initial levels. The peak in DNA content corresponds to an increase in the average number of nuclei per amoeba, following which the nuclear number decreases as the cells enter stationary growth phase. The RNA content is approximately 18 pg per amoeba (Weik and John, 1977; 1978; John, 1982).

6.5.1. Membrane carbohydrate moieties

N. fowleri trophozoites expresses high levels of surface glycoconjugates that contain alpha-D-mannose, alpha-D-glucose, and terminal alpha-L-fucose residues (González-Robles et al., (2007; Cervantes-Sandoval et al., 2010). Glycoconjugates that contain D-mannose and L-fucose residues participate in the adhesion of *N. fowleri* and subsequent damage to MDCK cells (Cervantes-Sandoval et al., 2010). Notably, Josephson et al., (1977) reported that concanavalin A (Con A) agglutinated *N. gruberi* but did not agglutinate *N. fowleri*. These data indicate a difference in polysaccharide structure of cell membranes of *N. fowleri* and *N. gruberi* (Josephson et al., 1977). Similarly, Bose et al., (1989) showed that during *in vitro* conversion from trophozoites to cysts, *N. fowleri* cysts gained WGA-specific saccharide and lost UEA I-specific saccharides suggesting that the surface of the organism exhibited replacement of saccharides during conversion (Bose et al., 1989).

6.5.2. Trypanothione/trypanothione reductase and glutathione/glutathione reductase systems

N. fowleri possesses the thiol compound trypanothione, which was previously thought to occur only in Kinetoplastida. The presence of the trypanothione/trypanothione reductase system in *N. fowleri* creates the possibility of using this enzyme as a new "drug target" for rationally designed drugs to eliminate the parasite, without affecting the human host (Ondarza et al., 2006).

6.5.3. Selenocysteine biosynthesis

Selenium (Se) is an essential trace element primarily found in selenoproteins as the 21st amino acid (selenocysteine, Sec, or U). Selenoproteins play an important role in growth and proliferation and are typically involved in cellular redox balance and has been identified in *Naegleria* (da Silva et al., 2013).

6.5.4. Expression of CD45-like glycoprotein

CD45 is a 180 - 240 kD single chain type I membrane glycoprotein. CD45 is a signaling molecule that regulates a variety of cellular processes including cell growth, differentiation, cell cycle, and oncogenic transformation. CD45 is known to play a critical role in T- and B-cell antigen receptor-mediated activation by dephosphorylating substrates including p56Lck, p59Fyn, and other Src family kinases. Ravine et al., (2010) showed that anti-human leukocyte antibody, CD45, demonstrated strong reactivity with *N. fowleri* suggesting expression of CD45 glycoprotein on the surface of *N. fowleri*. These results suggested a possible utility of using anti-human leukocyte antibody to screen *Naegleria* cells for similarly expressed protein epitopes.

6.5.5. Adenylyl cyclases

N. australiensis genome showed that the organism encodes polypeptides similar to photoactivated adenylyl cyclases. Each of the *Naegleria* proteins consists of a "sensors of blue-light using FAD" domain and an adenylyl cyclase domain (AC domain). Analysis of the *N. fowleri* genome revealed that the organism encodes a protein bearing an amino acid sequence similar to a "sensors of blue-light using FAD" domain. Experimental results indicated that the sequence similar to the BLUF domain found in *N. fowleri* functioned as a sensor of blue light (Yasukawa et al.,, 2013).

6.5.6. Beta-glucosidase and beta-galactosidase

N. fowleri grown axenically, contain high levels of beta-D-glucosidase. When the amoebae are subjected to freeze-thawing, sonication, and centrifugation (100,000 g, 1 h), 85% of the beta-glucosidase activity appears in the supernatant fraction. The predominant soluble beta-D-galactosidase activity in the *Naegleria* extract co-purifies with the beta-D-glucosidase; the two activities have the same isoelectric point (pI, 6.9), similar heat stabilities, are both inhibited by lactobionic acid (Ki, 0.40 mM), and exhibit optima at pH 4.5, indicating that they are probably the same enzyme. *Naegleria* beta-D-glucosidase has an apparent molecular weight of 66 kDa, a Stokes radius of 25 A, and a sedimentation coefficient of 4.2S. The beta-glucosidase is not inhibited by conduritol beta-epoxide or galactosylsphingosine but is completely inhibited by 1.25 mM bromo conduritol beta-epoxide. The latter compound, when present in the growth medium, inhibits the growth of the organism and profoundly alters its ultrastructure, the main effect being the apparent inhibition of cytokinesis and the generation of multinucleate cells (Das et al., 1987).

6.5.7. Acid phosphatase and heme proteins

The heme proteins constitute major structural elements in the *N. fowleri*. Strongly positive electron-dense globular bodies appear to be cytoplasmic storage sites for lysosomal hydrolases and/or catalase. It is proposed that *N. fowleri* actively utilize host erythrocytes as their major source of nutriment (Feldman, 1977). It was suggested that *N. fowleri*'s rapid invasive behavior is the physiologic result of the overwhelming availability of erythrocytes in the host inflammatory reaction, and the combined mechanisms of enhanced extracellular destruction with subsequent phagocytosis. Highly virulent strains of *N. fowleri* produced a magnitude more catalase, phospholipase A, and sphingo-myelinase activity than low-virulent strains (reviewed in John, 1982).

6.5.8. Pyrophosphate-dependent phosphofructokinase

PPi-dependent phosphofructokinase (PPi-PFK) have been detected in extracts of *N. fowleri*, with a specific activity of about 15-30 nmol/min mg of protein, which was increased about 2-fold by 0.5 mM AMP. PPi-PFK was inactivated upon gel filtration and could be re-activated by incubation at 30°C in the presence of AMP. The pure enzyme had a specific activity of 65 µmol/min mg of protein, and SDS/PAGE analysis showed a single band, of 51 kDa. Size-exclusion chromatography revealed the existence of two forms: a large one (approximately 180 kDa), presumably a tetramer, which was active, and a smaller one (approximately 45 kDa), presumably the monomer, which was inactive, but could be re-activated and converted into the large form by incubation at 30°C in the presence of 0.5 mM AMP. Inactivation of the tetrameric enzyme was promoted by 0.25 M potassium thiocyanate. The enzyme displayed Km values of 10 and 15 µM for fructose 6-phosphate and PPi, respectively, in the forward reaction, and of 35 and 590 µM for fructose 1,6-bisphosphate and Pi in the backward reaction. The activity was dependent on the presence of Mg^{2+}. AMP increased Vmax. about 2-fold without changing the affinity for the substrates; its half-maximal effect was observed at 2 µM (Mertens et al., 1993). The coding sequence of the cDNA consists of 1311 bases which translates into 437 amino acids with a molecular mass of 48 kDa. The activity was lost on incubation with the chaotropic agent, KSCN, and recovered by subsequent incubation with AMP. No nucleotide-binding motif or evidence for a nucleotide-binding site characteristic of the ATP-dependent phosphofructokinases would be found was the primary structure (Wessberg et al., 1995).

6.5.9. Cytosolic heat shock protein 70

A gene encoding a cytosolic heat shock protein 70 from *N. fowleri* (Nf-cHSP70) was identified. The Nf-cHSP70 was 2,062 bp in length with an open reading frame of 1,980 bp encoding 659 amino acid residues. The deduced amino acid sequence of the gene shared high sequence identities with HSP70s from other organisms and mammals. The characteristic domains, including *N*-terminal ATPase domain, calmodulin-binding domain, and EE(D)VD motif, found in HSP70s were also well conserved in this gene. The recombinant Nf-cHSP70 protein showed strong antigenicity against the sera from mice experimentally infected with *N. fowleri*. Immunofluorescence assay showed that Nf-cHSP70 localized in cytosol of the parasite. The expression levels of gene transcripts, and its products were significantly increased at high temperature (42°C). The definitive biological roles of Nf-cHSP70 are not clear, but it may protect the parasite under environmental changes especially high temperature (Song et al., 2007).

6.5.10. Low-molecular-mass thiol compounds

Low-molecular-mass thiol compounds in *N. fowleri* have been described (Ondarza et al., 2003). The amounts detected are expressed in nmol/1 x 10^6 trophozoites cultivated at various stages of growth in the appropriate culture medium. Unlike cysteine and glutathione, a number of these are not represented in normal human lymphocytes. Some of these thiol compounds from *N. fowleri* must have their respective disulphide reductases, although the presence of thiol-disulphide exchange reactions must be considered. The presence of thiol compounds in *N. fowleri* which are not present in human lymphocytes opens the possibility of searching for disulphide-reducing enzymes that can serve as drug targets (Ondarza et al., 2003).

6.5.11. Membrane-bound black bodies

Membrane-bound black bodies were found in all exponentially growing cell populations (Stevens et al., 1978). The bodies, 40 - 80 nm in diameter, were distributed randomly in the cytoplasm of *Naegleria* with ultrastructural features typical of trophozoites. No evidence was obtained that the contents of the black bodies were synthesized in the rough endoplasmic reticulum and packaged by membranous components, which could be a primitive "Golgi complex" in these amoebae. Examination of cells in various stages of encystation indicated that at least some of the cyst wall material was synthesized and packaged by the rough endoplasmic reticulum. After condensation into amorphous granules in the cisternae, the cyst wall material appeared in vesicles of the rough ER; these were frequently seen in close proximity to the cell membrane in the vicinity of developing cyst wall. Amorphous granules (~100 nm in diameter), which had variable densities and did not appear to be membrane bound, were seen in the cytoplasm of encysting cells. The substance of these granules also seemed to be incorporated into the cyst wall (Stevens et al., 1978). The membrane-bound black bodies appeared to be destroyed in lysosomal elements during encystation. The membrane-bound black bodies are suggested to be characteristic of trophozoites and unrelated to encystation of *Naegleria*.

6.5.12. Tet-like dioxygenase

Cytosine residues in mammalian DNA occur in five forms: cytosine (C), 5-methylcytosine (5mC), 5-hydroxymethyl-cytosine (5hmC), 5-formylcytosine (5fC) and 5-carboxylcytosine (5caC). The ten-eleven translocation (Tet) dioxygenases convert 5mC to 5hmC, 5fC and 5caC in three consecutive, Fe(II)- and α-ketoglutarate-dependent oxidation reactions. The Tet family of dioxygenases is widely distributed across the tree of life. The genome of *Naegleria* encodes homologues of mammalian DNA methyltransferase and Tet proteins. *Naegleria* Tet-like proteins (NgTet1), which shares significant sequence conservation (approximately 14% identity or 39% similarity) with mammalian Tet1. Like mammalian Tet proteins, NgTet1 acts on 5mC and generates 5hmC, 5fC and 5caC. The crystal structure of NgTet1 in complex with DNA containing a 5mCpG site revealed that NgTet1 uses a base-flipping mechanism to access 5mC. The

DNA is contacted from the minor groove and bent towards the major groove. The flipped 5mC is positioned in the active-site pocket with planar stacking contacts, Watson-Crick polar hydrogen bonds and van der Waals interactions specific for 5mC. The sequence conservation between NgTet1 and mammalian Tet1, including residues involved in structural integrity and functional significance, suggests structural conservation across phyla (Hashimoto et al., 2014).

6.5.13. Sterol biosynthesis

Cycloartenol, the sterol precursor in photosynthetic organisms, is present in *Naegleria*. It is accompanied by lanosterol and parkeol, as well as by the 24,25-dihydro derivatives of these triterpenes. One of the most striking features of *Naegleria* is the accumulation of 4 alpha-methylsterols which are present in similar amounts as those of 4,4-desmethylsterols (3-5 mg/g, dry weight). 4 alpha-Methylergosta-7,22-dienol was identified as a new compound. Ergosterol was the major 4,4-desmethylsterol, accompanied by small amounts of C27 and other C28 sterols. Treatment of *Naegleria* with fenpropimorph modified the sterol pattern of this amoeba and inhibited its growth. This fungicide, known to inhibit steps of sterol biosynthesis in fungi and plants, induced the disappearance of 4 alpha-methyl-delta 7-sterols and the appearance of the unusual delta 6,8,22-ergostatrienol. These results were explained by a partial inhibition of the delta 8-delta 7 isomerase, the small amounts of delta 7-sterols formed being converted into ergosterol which is still present in fenpropimorph-exposed cells. Cycloartenol, the sterol precursor in plants and algae, is also the sterol precursor in *Naegleria* species. Lanosterol, the sterol precursor in non-photosynthetic phyla (animal and fungi) and parkeol are more likely dead-ends of this biosynthetic pathway. The peculiar phylogenetic position of these protozoa was further emphasized by the action of indole acetic acid and other auxin-like compounds on their growth. Indeed amoebic growth was enhanced in the presence of these higher plant growth hormones. The differences in the sterol composition of the protozoa we have hitherto examined is related to their sensitivity toward polyene macrolide antibiotics (Raederstorff and Rohmer, 1987).

6.5.14. Other enzymes

Several other enzymes have been detected in *N. fowleri* including malic enzyme, 3-hydroxybutyrate dehydrogenase, isocitrate dehydrogenase, 6-phosphogluconate dehydrogenase, *L*-threonine dehydrogenase, glyceraldehyde-3-phosphate dehydrogenase, superoxide dismutase, hexokinase, phosphoglucomutase, uridine diphosphate-glucose pyrophosphorylase, 3-N-acetylglucosaminidase, aldolase, and glucose phosphate isomerase, proteases, lipases, phosphatases, esterases (Pernin et al., 1985).

6.6. Genome of the genus *Naegleria*

Originally, the number of chromosomes for *Naegleria* was suggested to be 3 - 16, however, later studies indicated that chromosomes were too small and too tightly packed to count (reviewed in Byers, 1986). The occurrence of complex banding patterns provided the first evidence of a diploid structure of the genome of these amoebae (Cariou and Pernin, 1987). The chromosome pattern of each species and subspecies was found to be distinct. Between 15 and 23 bands were resolved, with chromosome sizes ranging from a few hundred kilobases to about 1.5 Mb. Hybridisation with cloned rDNA identified one band in all species, corresponding to the rDNA plasmid that does not migrate according to its molecular weight because it is circular (De Jonckheere 1989; Clark et al., 1990).

The nuclear DNA content of *Naegleria* is circa 0.2 - 0.3 pg / amoeba, whereas *Acanthamoeba* has 1 - 2 pg / amoeba, while *Amoeba proteus* has about 34 - 43 pg / amoeba (Byers, 1986). DNA content was relatively constant during log to post-log phase transition of *N. fowleri* (Weik and John, 1978). Both nuclear and mitochondrial DNA contents vary during culture growth (Byers, 1986). *N. fowleri* genome is diploid (66 MB), while the haploid genome size is 29.62 MB with GC content of 35.4% and 17,252 open reading frames (Zysset-Burri et al., 2014). Of interest, the haploid genome size of *N. gruberi* is 40.96 MB with GC content of 33.1%, and composed of at least 12 chromosomes with sizes ranging from 0.7 to 6.5 MB. Additionally, it contains a 14 kb extrachromosomal plasmid and a 50 kb mitochondrial genome (Fritz-Laylin et al., 2011; Zysset-Burri et al., 2014).

6.6.1. The mitochondrial genome and a 60-kb nuclear DNA segment

In *N. gruberi*, mitochondrial DNA was estimated at 14% of the total cell DNA (Fulton, 1977), while *A. castellanii* contained 20 - 30% of the total cell DNA (Byers, 1986). The mitochondrial genome and a 60-kb segment of nuclear genome from *N. fowleri* is sequenced. The mitochondrial genome is highly similar to its counterpart in *N. gruberi* in gene complement and organization, while distinct lack of synteny is observed for the nuclear segments. Even in this short (60-kb) segment, potential factors for pathogenesis have been identified, including ten novel *N. fowleri*-specific genes. In addition, a homolog of cathepsin B; proteases proposed to be involved in the pathogenesis of diverse eukaryotic pathogens, including *N. fowleri* was identified (Herman et al., 2013).

6.7. Mitochondrial RNA editing

RNA editing converts hundreds of cytidines into uridines. Recognition of the RNA editing sites in the organelle transcriptomes requires numerous specific, nuclear-encoded RNA-binding pentatricopeptide repeat (PPR) proteins with characteristic carboxy-terminal protein domain extensions (E/DYW) previously thought to be unique to plants. However, a small gene family of such plant-like PPR proteins of the

DYW-type was recently discovered in the genome of *N. gruberi*. This raised the possibility that plant-like RNA editing may occur in *Naegleria*. A particular type of pentatricopeptide repeat (PPR) proteins with variable length of the 35 aa PPR motifs and conserved carboxyterminal extensions, named the PLS proteins was identified. Knoop and Rüdinger (2010) reported the first case of DYW-type PLS proteins outside of the plant kingdom in *N. gruberi* and hypothesize on horizontal gene transfer in very early land plant evolution. The mitochondrial transcriptome of *N. gruberi* was investigated and the identification of two sites of C-to-U RNA editing in the *cox1* gene and in the *cox3* gene were reported, both of which reconstitute amino acid codon identities highly conserved in evolution. An estimated 1.5 billion years of evolution separate the heterolobosean protist *Naegleria* from the plant lineage. The new findings either suggest horizontal gene transfer of RNA editing factors or that plant-type RNA editing is evolutionarily much more ancestral than previously thought and yet to be discovered in many other ancient eukaryotic lineages (Rüdinger et al., 2011).

6.8. RNA polymerase

Naegleria contains three major RNA polymerase activities (Soll and Fulton, 1974). One is resistant to the drug, α-amanitin and the other two are sensitive to it.

6.9. Ribosomal DNA (rDNA)

The rRNA genes are transcribed together in the following order: small subunit (18S) rDNA, an internal transcribed spacer (ITS1), 5.8SrDNA, a second ITS (ITS2), and the large subunit (28S) rDNA. In most species, the ITS1 is about 33 - 41 nucleotides in length and the ITS2 about 100 - 115 (Fritz-Laylin et al., 2011). The rDNAs of *N. gruberi* (28S, 18S, 5.8S) are encoded on an extrachromosomal circular nucleolar plasmid carrying all three rDNA genes. The 3,000 to 5,000 copies per cell of this 14-kilobase-pair circular plasmid carry all the 18S, 28S, and 5.8S rRNA genes. The presence of the ribosomal DNA of an organism exclusively on a circular extrachromosomal element is rare (Clark and Cross, 1987). Comparison of the small-subunit rRNA gene with the rRNA sequences of other eukaryotes resulted in a phylogenetic tree that supports the suggested polyphyletic origin of amoebae and suggests a flagellate ancestry for *Naegleria* (Clark and Cross, 1988; Baverstock et al., 1989). The entire 14,128 bp of the extrachromosomal circular rDNA plasmid was sequenced. It contained a single rRNA gene unit composed of 18S, 5.8S, and 28S rRNA genes, but no tRNA or 5S RNA genes. It was predicted that there are two open reading frames. The region that flanks the rRNA gene unit is A/T-rich, except for a highly G/C-rich region that is approximately 900 bp upstream of the rRNA genes. Fluorescence *in situ* hybridization revealed that the rDNA plasmids cluster within the nucleolus, suggesting that they are highly organized for the efficient transcription of rRNAs. The *N. gruberi* rDNA plasmid has a unique high-order cluster structure that provides both a molecular basis for understanding chromosomal organization in basal eukaryotes, and a vehicle for constructing stable transgenic vectors (Maruyama and Nozaki, 2007).

6.9.1. Large subunit ribosomal DNA

The group I introns in the large subunit ribosomal DNA (LSUrDNA) in *Naegleria* are at different locations and are probably acquired by horizontal transfer, contrary to the SSUrDNA group I introns in this genus which are of ancestral origin and are transmitted vertically (De Jonckheere and Brown, 1998). The two group I introns Nae.L1926 and Nmo.L2563, found at two different sites in nuclear LSU rRNA genes of *Naegleria*. Nae.L1926, and Nmo.L2563 RNAs self-splice *in vitro*, generating ligated exons and full-length intron circles as well as internal processed excised intron RNAs. Formation of full-length intron circles is found to be a general feature in RNA processing of ORF-containing nuclear group I introns. Both *Naegleria* LSU rDNA introns contain a conserved polyadenylation signal at exactly the same position in the 3' end of the ORFs close to the internal processing sites, indicating an RNA polymerase II-like expression pathway of intron proteins *in vivo*. The intron proteins I-NaeI and I-NmoI encoded by Nae.L1926 and Nmo.L2563, respectively, correspond to His-Cys homing endonucleases of 148 and 175 amino acids (Haugen et al., 2002).

6.9.2. Small subunit ribosomal DNA

The nuclear small subunit ribosomal RNA (SSU rRNA) gene harbors an optional group I intron. The intron (Nae.S516) has a complex organization of two ribozyme domains (NaGIR1 and NaGIR2) and a homing endonuclease gene (NaHEG). NaGIR2 is responsible for intron excision, exon ligation, and full-length intron RNA circularization, reactions typical for nuclear group I intron ribozymes. NaGIR1, however, is essential for NaHEG expression by generating the 5' end of the homing endonuclease mRNA. The structural organization and catalytic properties of the large nucleolar group I introns (NaSSU1) were characterized. NaSSU1 consists of three distinct RNA domains: an open reading frame encoding a homing-type endonuclease, and a small group I ribozyme (NaGIR1) inserted into the P6 loop of a second group I ribozyme (NaGIR2). The two ribozymes have different functions in RNA splicing and processing. NaGIR1 is an unusual self-cleaving group I ribozyme responsible for intron processing at two internal sites (IPS1 and IPS2), both close to the 5' end of the open reading frame. This processing is hypothesized to lead to formation of a messenger RNA for the endonuclease. Structurally, NaGIR2 is a typical group IC1 ribozyme, catalyzing intron excision and exon ligation reactions. NaGIR2 is responsible for circularization of the excised intron, a reaction that generates full-length RNA circles of wild-type intron. These twin-ribozyme introns define a distinct category of group I introns with a conserved structural organization and function (Einvik et al., 1997).

Later studies suggested that the NaGIR1-catalyzed self-cleavage of the intron RNA, which is a key event in expression of the endonuclease (Decatur et al., 2000). Divalent metal ions, primarily Mg^{2+}, are essential for *Naegleria* endonuclease activity. Whereas both Mn^{2+} and Ca^{2+} could substitute for Mg^{2+}, but with a slower cleavage rate, Zn^{2+} was unable to support cleavage. Interestingly, the pH dependence of DNA cleavage was found to vary significantly between the I-NitI and I-NjaI/I-NanI endonucleases with optimal pH values of 6.5 and 9, respectively. Site-directed mutagenesis of conserved I-NjaI residues strongly supports the hypothesis that *Naegleria* homing endonucleases share a similar zinc-binding structure and active site with the His-Cys box homing endonuclease I-PpoI (Elde et al., 1999; 2000).

6.9.3. Kinetic and secondary structure analysis of group I ribozyme

NanGIR1 is a catalytic element inserted in the P6 loop of a group I intron (NanGIR2) in the small subunit rRNA precursor. It catalyzes site-specific hydrolysis at an internal processing site (IPS) after a G residue that immediately follows the P9 stem-loop. Functional and structural analyses were initiated to compare NanGIR1 to group I introns that carry out self-splicing. Chemical modification and site-directed mutagenesis studies showed that NanGIR1 shares many structural elements with other group I introns, but also contains a pseudoknot (P15), which is important for catalytic activity. Deletion analysis revealed the boundaries of the minimum self-cleaving unit (178 nucleotides). The rate of self-cleavage was measured as a function of mono- and divalent ion concentration, temperature, and pH. The reaction at the IPS yields 5'-phosphate and 3'-hydroxyl termini, requires Mg^{2+}or Mn^{2+} ions, and is first-order in $[OH^{-}]$ between pH 5.0 and 8.5. The latter results suggest that the nucleophile in the reaction is hydroxide or possibly a Mg^{2+} coordinated hydroxide (Jabri et al., 1997).

Later studies show the presence of a new category of self-splicing group I introns with conserved structural organization and function. These complex rDNA introns contain two distinct ribozymes with different functions: a regular group I splicing-ribozyme and a small internal group I-like ribozyme (GIR1), probably involved in protein expression. GIR1 was found to cleave at two internal sites in an obligate sequential order. Both sites are located 3' of the catalytic core. GIR1-catalyzed transesterification reactions could not be detected. GIR1 lack most peripheral sequence components, as well as a P1 segment, and, at approximately 160-190 nt, they are the smallest functional group I ribozymes known from nature. GIR1 was found to contain a novel 6-bp pseudoknot (P15) within their catalytic core region (Einvik et al., 1998). The Nae.S516 twin-ribozyme intron was gained early in the *Naegleria* evolution with subsequent vertical inheritance. The intron was lost in the majority of isolates (70%), leaving a widespread but scattered distribution pattern. Why the apparent asexual *Naegleria* amoebae harbors an active intron homing endonuclease that is normally dependent on sexual reproduction for its function requires further investigation (Wikmark et al., 2006).

6.10. Classification

The International Society of Protistologists has classified Eukaryotes into six "Super Groups" namely, Amoebozoa, Opisthokonta, Rhizaria, Archaeplastida, Chromalveolata, and Excavata. Among free-living pathogens, genus *Acanthamoeba* and *Balamuthia* are included under Super Group Amoebozoa: Acanthamoebidae; while genus *Sappinia* under Super Group Amoebozoa: Flabellinea: Thecamoebidae; and genus *Naegleria* under Super Group Excavata: Heterolobosia: Vahlkampfiidae (Adl et al., 2005).

Naegleria fowleri belong to the genus *Naegleria*, family Vahlkampfiidae in the class Heterolobosea. Like all other members of this class, it is a free-living protist that feeds mostly on bacteria. Genus *Naegleria* contains both pathogenic and non-pathogenic species.

The rRNA genes have been used for classification. The rRNA genes are transcribed together in the following order: 18S rDNA, an internal transcribed spacer (ITS1), 5.8S rDNA, ITS2, and 28S rDNA. In most species, ITS1 is about 33 - 41 nucleotides in length and the ITS2 about 100 - 115. The ITS sequence is rapidly evolving and it has been used to classify over 40 species of *Naegleria* (DeJonckheere, 2004). However, *N. fowleri* is the only species responsible for producing primary amoebic meningoencephalitis infection in humans. Like other pathogenic microbes, it remained a challenge to differentiate pathogenic and non-pathogenic *Naegleria*. Several methods have been employed as described below. At first *N. fowleri* isolates were identified by pathogenicity test in mice combined with testing the growth temperature at a maximum of 45°C. *N. fowleri* can grow at temperatures up to 45°C. In routine testing of samples for *N. fowleri*, samples are incubated between 42°C - 45°C in an attempt to suppress the growth of other amoebae. Later antibodies and isoenzyme techniques were used to identify *N. fowleri*. Restriction-fragment-length polymorphism (RFLP) (De Jonckheere, 1987, 1988) and electrophoretic karyotyping (KE) (De Jonckheere, 1989) have been used for typing *Naegleria* spp. and different strains of *N. fowleri*. It was discovered that strains of *N. fowleri* from Europe, the USA, and Australia and New Zealand differ in RFLP and KE, as well as in isoenzymes (De Jonckheere, 1988). However, the major advance in typing *N. fowleri* isolates, and identifying different *Naegleria* spp. (De Jonckheere, 1989) and other Vahlkampfiidae (De Jonckheere and Brown, 2005), came with determining the ITS1, 5.8S rDNA and ITS2 sequences.

The availability of molecular tools has made it possible to differentiate this genus into eight different clusters, and *N.*

fowleri belongs to cluster 8 (reviewed in De Jonckheere, 2011). Based on the genomic information, it is suggested that *N. fowleri* probably evolved in the USA from the non-pathogenic *N. lovaniensis* (also member of cluster 8), from where it spread worldwide (reviewed in De Jonckheere, 2011). *N. lovaniensis* is also the closest relative of *N. fowleri*. Interestingly, two species of *Naegleria* including *N. australiensis* and *N. italica* that have shown to produce infections in experimental mice are not closely related to *N. fowleri*. Both of these species belong to the cluster 5 of the Genus *Naegleria*. Furthermore, various strains within the species, *N. fowleri*, have been differentiated into eight types, based on the length of the internal transcribed spacer 1 (ITS1) and a one bp transition in the 5.8S rDNA.

The differences in ITS1 do allow to distinguish different *N. fowleri* isolates. The ITS and 5.8S rDNA sequences will be of additional help in describing new *Naegleria* spp. (De Jonckheere, 1998). Later studies tested one isolate each of *N. australiensis*, *N. gruberi*, *N. jadini*, and *N. lovaniensis* and 22 isolates of *N. fowleri* and characterized the ITS and mitochondrial small subunit rRNA (mtSSU rRNA) gene. The mtSSU rRNA primers amplified DNA of all isolates, with distinct sequences obtained from all species examined. In contrast, the ITS primers only amplified DNA from *N. lovaniensis* and *N. fowleri*, with minor sequence differences between the two. Three genotypes of *N. fowleri* were found among the isolates analyzed in both the mtSSU rRNA gene and ITS. The extent of sequence variation was greater in the mtSSU rRNA gene, but the ITS had the advantage of length polymorphism. These data should be useful in the development of molecular tools for rapid species differentiation and genotyping of *Naegleria* spp. (Zhou et al., 2003). Subsequent studies analyzed the ITS of four isolates of *N. fowleri*. Three strains were isolated from patients and one from the environment. All four strains were confirmed to be *N. fowleri* by species-specific PCR. The ITS lengths observed were ITS1, 85 bp; ITS2, 106 bp; and 5.8S, 176 bp. The ITS main products of the strain from the environment were similar to those of the clinical strains (Tiewcharoen et al., 2007).

7

Cellular Differentiation in *N. fowleri*

Abstract
N. fowleri has a fascinating biology and exhibits three life forms. It reproduces in the amoeboid form, transforms into a flagellate to travel long distance to search for food, and switches into a dormant cyst to withstand harsh conditions. As a flagellate, it is a useful model organism to study motility processes of flagellates; as an amoeba, it is a useful model to study molecular biology of phagocytosis; and its ability to transform into cysts that remain viable for months to years, makes it an attractive model to study cellular differentiation processes of cell dormancy. In this chapter, we discuss the present knowledge of cellular differentiation processes in *N. fowleri* and associated factors.

7.1. Cellular differentiation

Naegleria fowleri exhibit three stages, amoeboid stage, cyst stage and a flagellar stage (Figure 7.1). The amoeboid form moves by forming one or more lobose pseudopodia and show progressive flowing movement. Trophozoites of *N. fowleri* are long and slender (8-15 µm). The anterior pseudopodal end have a higher negative surface charge than the rest of the cell (Forrester et al., 1967; Antonios, 2010). Scanning electron microscopy revealed that the surface of trophozoites appeared undulating, wrinkled and covered at irregular intervals by protruding vesicles (Antonios, 2010). *N. fowleri* exhibited surface extensions which are long and thin. These extensions are suggested to aid in amoebae contact with and cytolysis of host cells. In the absence of food but presence of water, the amoeboid form develops flagella (Fulton and Dingle, 1967). The capacity of *Naegleria* to transform is stably inherited, and the phenotypic changes occur without change in genotype (Fulton and Dingle, 1967). The number of flagella per flagellate stage varies from 1 to 8, with 2.4 being the average number per cell. John et al., (1991) show that the majority of *N. fowleri* cells had 2 flagella. However, the individual response of cells is extremely heterogeneous with some developing no more than the normal 2 flagella, whereas others develop as many

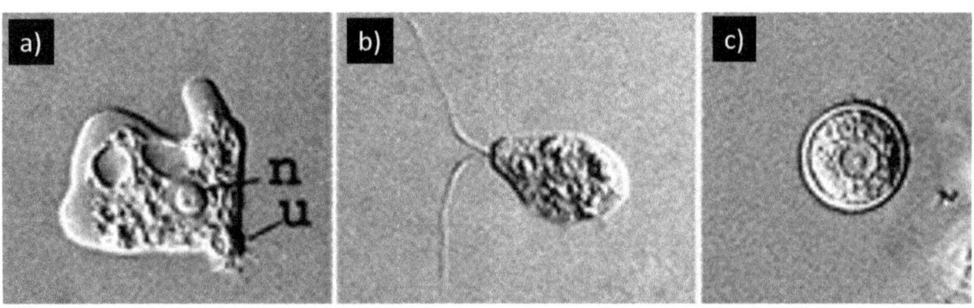

Figure 7.1. Under favourable condition, *N. fowleri* exhibits trophozoite form (a), and in the absence of food but presence of water, it transforms into a flagellate form (b). Under harsh conditions, trophozoite transforms into a cyst form (c). Viewed under phase contrast microscopy. X1000. (courtesy: G. S. Visvesvara, CDC, USA).

as over 10. After approximately 1 h, the flagellates revert to amoebae and loose flagella (Dingle, 1970). When subjected to high temperatures during flagellate differentiation, populations develop an average of 4-5 flagella per cell (Dingle, 1979). Attempts to maximize this phenomenon by altering cellular and environmental variables revealed that (i) a few isolates become multiflagellated, (ii) temperature is the most critical variable as the highest numbers of flagella are obtained at high temperature, although pH alone does not affect numbers of flagella, (iii) a pH optimum of 5.5 - 7.0 exists for temperature-shocked cells, (iv) single cells in microdrops become multiflagellated, but the population response is density-dependent (Dingle, 1979). Basal bodies are constructed during transformation and, in cells transforming synchronously at 25°C, they are first seen about 10 min before flagella are seen. No structural precursor for these basal bodies has been found (Fulton and Dingle, 1971).

Under harsh conditions, amoebae transform into cysts. Cysts are varied in morphological appearance with some appearing wrinkled while others smooth. Additionally, empty cysts are also seen with many pores on the surface (Antonios 2010). Saanen (1980) showed that cysts can be smooth (*N. fowleri*) or rough (*N. gruberi*). The average number of pores varied between 1.2 and 7.2. Their margin can be smooth (*N. fowleri*) or pierced (*N. gruberi*). The thickness of the bordering cell wall varied between 0.4 and 1.0μm and that of the pore is about 0.6μm (Saanen, 1980).

7.2. Proteins in flagellates and growing amoebae of *N. fowleri*

Using two-dimensional polyacrylamide gel electrophoresis, the majority of the polypeptides of amoebae and flagellates exhibit molecular sizes in the range of 20 to 60 kDa. The greatest number of polypeptides detected in amoebae and flagellates was in the isoelectric focusing range of pH 6 to 7.

The radioactivity per polypeptide species in the isoelectric focusing gradient below 6.3 was greater in amoebae than in flagellates. Polypeptides in the size range of 20 to 60 kDa had a median isoelectric point below pI 6.3, whereas those larger than 60 kDa had a median pI value above 6.3 (Woodworth et al., 1982).

7.3. Encystation: Amoeba to Cyst and *vice versa*

Highly pathogenic amoebae encyst rapidly during late stationary phase (Marciano-Cabral and Cline, 1987). Encystation is characterized by an increase in cytoplasmic density due to dehydration (Schuster, 1963b). The mitochondria are found closely associated with the endoplasmic reticulum. Numerous vacuoles (or droplets) of low electron density occurred in proximity to the nucleus and scattered throughout the cytoplasm. These were thought to contain the material which gives rise to the cyst wall. Other types of vacuoles are also seen. The mature cyst wall is double, consisting of an inner thick component (200 - 450 μm) and an outer thin component (25 μm). The two layers are separated by a space filled with a spongy network but joined at the region of the cyst pore. The pores (about 600 μm) are closed by a plug of electron-transparent material. At excystation, the amoeba cytoplasm becomes highly alveolar. The plug sealing the pore in the wall dissolves and the amoeba leaves the cyst (Schuster 1963b). During *in vitro* conversion from trophozoites to cysts, *N. fowleri* cysts gain WGA-specific saccharide and lose UEA I-specific saccharides (Antonios, 2010).

7.4. Ultrastructural study of the encystation and excystation processes

In the initial stages of encystation, the cisternae of the endoplasmic reticulum becomes densely filled with a fibrillar material. Vesicles with a similar content that appears to be derived from the cisternae is also observed in close contact with the plasma membrane. As encystation progresses, the fibrillar material becomes localized on the surface of the amoeba. An irregular compaction is observed in some areas of the cyst wall, which contains thin extensions of the cyst wall fibrillar material. Completely formed cysts had two to three ostioles, each sealed by an operculum. The operculum contained two areas in which a differential compaction of the fibrillar structure is observed. When excystation is induced, small dense granules, which are in close contact with fibrillar material are observed in the cyst cytoplasm and in the peritrophic space. During excystation, the more compact component of the operculum moves to enable the pseudopod of the emerging trophozoite to penetrate the ostiole. Vacuoles containing a fibrillar material, probably derived from the cyst wall, are observed in the cytoplasm of the pseudopodia (Chávez-Munguía et al., 2009; Lastovica, 1974).

7.4.1. Effect of CO_2 on excystation

The exposure of cysts to slightly increased environmental CO_2 causes excystation (Averner and Fulton, 1966). It is suggested that the molecular CO_2 is an excellent signal to induce excystation. The optimum temperature for excystation is about 30°C and the optimum pCO_2 in air is 5%, while higher concentrations inhibited. Inhibitors of carbonic anhydrase reduced excystation. The excystation is inhibited by actinomycin D, and DNA transcription is apparently obligatory. Once excystation initiates, it proceeds to completion in atmospheric CO_2.

7.4.2. Effect of steroids

Cellular differentiation is affected by the presence of steroids in the medium (Pearson and Willmer, 1963). At high concentrations, progesterone and deoxycorticosterone prevent amoebae from changing to the flagellate form. When applied to the flagellate, progesterone show little action, but deoxycorticosterone encourages the return to the amoeboid form, especially in the presence of K^+. In lower concentrations both steroids favour the transformation. While the dose/response curves indicate qualitatively different effects of concentration of progesterone and deoxy-corticosterone, they only show quantitatively different effects with oestradiol and androsterone. Progesterone acts on amoebae at concentr-ations which are comparable with those at which it acted in the human body. When both progesterone and oestradiol are applied together at concentrations which suppressed the flagellate form, the effects are additive. The same applied to mixtures of testosterone and oestradiol (Pearson and Willmer, 1963).

7.4.3. Enolase is expressed during cyst differentiation

Cysts of *N. fowleri* present an external single-layered cyst wall. A monoclonal antibody (B4F2 mAb) that specifically recognizes enolase in the cyst wall of *Entamoeba invadens* recognized in soluble extracts of *N. fowleri* cysts, a 48 kDa protein with similar molecular weight to the enolase reported in *E. invadens* cysts. Immunofluorescence with the B4F2 mAb revealed positive cytoplasmic vesicles in encysting amoebae, as well as a positive reaction at the cell wall of mature cysts. Immunoelectron microscopy using the same monoclonal antibody confirmed the presence of enolase in the cell wall of *N. fowleri* cysts and in cytoplasmic vesicular structures. In addition, the B4F2 mAb had an inhibitory effect on encystation of *N. fowleri* (Chávez-Munguía et al., 2011). The enolase is known to be involved in glycolysis but several other glycolytic enzymes have structural roles in other protists. For example, aldolase is essential for energy production as well as play structural roles such as bridging adhesin-actin cytoskeletal interactions during apicomplexan parasite (*Toxoplasma gondii*) invasion of host cells (Starnes et al., 2009).

7.5. Flagellation: Amoebae to Flagellates

Flagella are formed by filamentous extension from endoplasmic protrusion (Chang, 1958b). One, two or three pairs of flagella are formed from a single protrusion. Occasionally a second protrusion gave rise to the 3rd pair of flagella. A chromatin body of cytoplasmic origin is always present at the base of the protrusion or of the flagella, and is believed to be the parabasal body. Occasionally one chromatin body is seen at the base of each pair of flagella. Alternating bands of light and dark areas are seen in the flagella soon after they are formed and persisted throughout the flagellate stage. Reversion from flagellate to amoeba stage is accomplished by absorption of the flagella, the shedding of one or more flagella and the absorption of the rest, or by casting-off a small part of the body to which the flagella are attached. Around 35% of the trophozoites underwent flagellate transformation in water when they were in metacystic stage or were harvested from plate cultures up to 12 h of incubation at 25 - 27°C; as the cultural age advanced, the percentage amoebae undergoing transformation became progressively lower. Prolonged cultivation in a liquid medium at 25 - 27°C, or on an agar medium at 37°C, resulted in temporary loss of ability to form flagellates. The freezing of trophozoites up to 40 min and of cysts up to 1 week at - 25°C to -30°C stimulated flagellate transformation. An increase in water temperature from 25 - 27°C to 37°C or decrease to 8 - 10°C suppressed the transformation, but after the flagellates were formed, the former temperature change shortened while the latter lengthened duration of the flagellate stage. Freezing of the amoebae in the flagellate stage at -25°C to -30°C exerted no adverse effect on their return to amoeboid form when the freezing time did not exceed 30 min. Cultures developed from single organisms isolated in the flagellate stage yielded amoebae showing flagellate transformation characteristics similar to those yielded by ordinary cultures (Chang, 1958b).

Flagellates have an interconnected flagellar apparatus, consisting of nucleus, rhizoplast and accessory filaments, basal bodies, and flagella (Dingle and Fulton, 1966). The development of the flagellar apparatus showed no indications of rhizoplast, basal body, or flagellum structures in amoebae. A basal body appears and assumes a position at the cell surface with its filaments perpendicular to the cell membrane. Axoneme filaments extend from the basal body filaments into a progressive evagination of the cell membrane which becomes the flagellum sheath. Continued elongation of the axoneme filaments leads to differentiation of a fully formed flagellum within 10 min after the appearance of basal bodies (Dingle and Fulton, 1966).

Amoebae from the stationary phase of growth enflagellate more readily than actively growing amoebae. Incubation in expended culture medium from stationary-phase cultures enhances the capability of growing amoebae to enflagellate after subculture to amoebasaline. Enflagellation is more extensive when the population density in amoebasaline did not exceed 2 x 10^5 amoebae per mL.

7.5.1. Effects of oxidative phosphorylation, protein synthesis, RNA synthesis, DNA synthesis on tranformation

Inhibitors of oxidative phosphorylation, protein synthesis, RNA synthesis, and DNA synthesis delayed or blocked the transformation, suggesting that RNA and protein synthesis are required (Yuyama, 1971) as expected. The differentiation is dependent on transcription. Net synthesis of RNA continues for the first 30 min after differentiation is initiated, but thereafter degradation exceeds synthesis, leading to an overall decrease of 6% in total RNA per cell by 90 min. [^{14}C]-Uracil is incorporated into RNA during differentiation. Incorporation is linear until roughly 10 min before the appearance of flagella, at which time the rate of incorporation declines. There is proportionally more synthesis of messenger-like RNA and less of rRNA in differentiating cells. The RNA is first present in the nucleoplasm, and moves rapidly to the cytoplasm, reaching equilibrium (nucleus/cytoplasm) within 10 min. In the cytoplasm it is found associated with polysomes, where it has the EDTA and ribonuclease sensitivity expected of mRNA (Walsh and Fulton, 1973).

Actinomycin D and daunomycin (inhibitors of RNA synthesis), and cycloheximide (inhibitor of protein synthesis), prevent differentiation if added soon after the cells are transferred to non-nutrient buffer but cease to block specific differentiation events (Fulton and Walsh, 1980). Cycloheximide at 0.5 µg per mL and actinomycin D at 25 µg per mL completely prevented enflagellation when added at time zero. Cycloheximide at 0.5 µg per mL, added 120 to 300 min after initiation of enflagellation, prevented further differentiation and caused existing flagellates to revert to amoeboid cells. Similarly, actinomycin D at 25 µg per mL, added 90 to 300 min after initiation of enflagellation, retarded differentiation and caused flagellates to revert. Radiolabeled precursors were incorporated into macromolecules during differentiation in non-nutrient buffer (Cable and John, 1986). After transition point, morphogenesis can occur, even in the presence of the inhibitor and in the virtual absence of transcription or translation (Fulton and Walsh, 1980). The differentiation in *Naegleria* involves a redirection of cell metabolism to produce new RNA and protein molecules that are essential for morphogenesis. Specific new proteins, including the tubulins that form the flagellar microtubules, are synthesized at various times during differentiation, and particular mRNA species appear and disappear. The time course of the synthesis of the alpha and beta subunits of flagellar tubulin is paralleled by the programmed appearance and disappearance of flagellar tubulin mRNAs. The evidence supports the hypothesis that the synthesis of flagellar tubulin is regulated by the transcription, and subsequent disappearance, of flagellar tubulin mRNA. Translatable mRNAs for two calmodulin-like calcium-binding proteins appear and disappear contemporaneously with those for flagellar tubulin. During differentiation, the synthesis of actin, the major protein of amoebae, is selectively shut down, and translatable actin mRNA rapidly disappears (Fulton 1983). Although flagellation is dependent on RNA synthesis, no detectable changes occurred qualitatively or quantitatively in RNA polymerases during differentiation. It was concluded that gross changes in RNA synthesis as well as differential gene activity can occur with absolutely no major fluctuations in the DNA-dependent RNA polymerases (Soll and Fulton, 1974).

Flagellation is accompanied with a decrease in DNA synthesis. Flagellation formation begins 60 min after incubation in transformation medium. The nuclear DNA synthesis decreased, while mitochondrial DNA synthesis continued and became increasingly dominant as differentiation progressed. The reduction in the nuclear DNA synthesis in differentiating cells is not due to the inhibition of initiation of DNA replication, but rather to the termination of the DNA replicating process. The DNA synthesis is curtailed in the presence of RNA and protein synthesis which are required for differentiation (Corff and Yuyama, 1976). The decrease in thymidine incorporation was accompanied with a drop in thymidine uptake as well as in the activity of thymidine kinase. In contrast, nucleotide phosphotransferase which phosphorylates thymidine to dTMP remained unchanged during flagellation. Unlike most protists that arrest in G1 or G2 phase during differentiation, *Naegleria* arrests in mid-S phase during flagellum formation (reviewed in Byers, 1986).

7.5.2. *De novo* formation of cytoplasmic cytoskeleton

When stressed, *Naegleria* rapidly (and synchronously) differentiates into a flagellate, forming a complete cytoplasmic cytoskeleton *de novo*, including two basal bodies and flagella. *Naegleria* has genes encoding conserved centriole proteins. Using novel antibodies, the localization of three centrosomal protein homologs (SAS-6, gamma-tubulin, and centrin-1) during the assembly of the flagellate microtubule cytoskeleton were described. *Naegleria* expresses the proteins in the order as follows: basal bodies, followed by centrin and finally gamma-tubulin. The similarities between basal body assembly in *Naegleria* and centriole assembly in animals indicate that mechanisms of assembly, as well as structure, have been conserved throughout eukaryotic evolution (Fritz-Laylin et al., 2010).

The *de novo* formation of basal bodies in *Naegleria* is preceded by the transient formation of a microtubule-nucleating complex containing gamma-tubulin, pericentrin, and myosin II complex (GPM complex). The microtubule-nucleating activity of GPM complexes is maximal just before the formation of visible basal bodies and then rapidly decreased. The regulation of microtubule-nucleating activity of GPM complexes is accomplished by a transient phosphorylation of the complex. Inhibition of dephosphorylation after the formation of basal bodies resulted in the formation of multiple flagella. These data suggest that the nucleation of microtubules by GPM complexes precedes the *de novo* formation of basal bodies and that the regulation of microtubule-nucleating activity of GPM complexes is essential to the regulation of basal body number (Kim et al., 2005).

7.5.3. Synthesis and assembly of the cytoskeleton of flagellates

When flagellates are extracted with non-ionic detergent and stained by the indirect immunofluorescence method with AA-4.3 (a monoclonal antibody against *Naegleria* beta-tubulin), flagella and a network of cytoskeletal microtubules (CSMT) are seen. When *Naegleria* amoebae are examined in the same way, no cytoplasmic tubulin-containing structures are seen (Walsh, 1984). Formation of the flagellate cytoskeleton is followed during the differentiation of amoebae into flagellates by staining cells with AA-4.3. The first tubulin containing structures are a few cytoplasmic microtubules that form at the time amoebae rounded up into spherical cells. The formation of these microtubules is followed by the appearance of basal bodies and flagella and then by the formation of the CSMT. The CSMT formed before the cells assumes the flagellate shape. In flagellate shaped cells, the CSMT radiate from the base of the flagella and follow a curving path the full length of the cell. Protein synthetic requirements for the formation of CSMT are examined by transferring cells to cycloheximide at various times after initiation. One-half the population completed the protein synthesis essential for formation of CSMT, 61 min after initiation of the differentiation. This is 10 min after the time when protein synthesis for formation of flagella is completed and 10-15 min before the time, when the protein synthesis necessary for formation of the flagellate shape is completed (Walsh, 1984)..

7.5.4. Flagellar rootlet during flagellate differentiation

The time and sequence of rootlet development have been described in differentiating cells, and related to the appearance of basal bodies and flagella and to the assembly of the mature, functional flagellar apparatus. Developing rootlets, which cannot be identified with certainty until they are 1.5 - 2 µm long, are seen in the earliest transforming cells ~65 min post-initiation. Kinetics of rootlet appearance parallel those of basal bodies and flagella, with flagella preceding the rootlets by about 5 min (T_{50} for developing rootlets is 76 min). However, a brightly fluorescing condensate of rootlet protein appears in the cytoplasm at the edge of the nucleus at 58 min. This presumed initiating center for rootlet assembly thus condenses in the vicinity of, and virtually simultaneously with, the flagellar basal bodies. Rootlets attain their mature length (~13 µm) within about 30 min of their initial appearance. They then disappear completely from cells as flagella are lost during reversion to amoebae. Subsequently rootlets are reassembled simultaneously with the reappearance of flagella in cells undergoing a second round of flagellate differentiation (Larson and Dingle, 1981b).

7.5.5. Synthesis of centriole and flagella proteins

Centrin is an integral component of the basal bodies. In many organisms, centrin appears to be a constitutive protein, but in *Naegleria*, centrin gene expression occurs only during differentiation. Centrin mRNA, which has not been detected in amoebae, appears and disappears earlier in differentiation than a coordinately regulated set of differentiation-specific mRNAs encoding flagellar tubulin and calmodulin. Centrin antigen accumulates during differentiation, and then decreases in abundance as the flagellates mature and revert to amoebae. No localization of centrin has been detected in amoebae. During differentiation, centrin becomes localized to the basal bodies as soon as these structures are detected with anti-tubulin antibodies, first as a single dot and finally as two basal bodies. During reversion of flagellates to amoebae, centrin remains localized to the basal bodies for as long as they are present. When assembly of tubulin-containing structures during differentiation is prevented using oryzalin, centrin localization is prevented as well, yet inhibition of assembly does not affect accumulation of centrin antigen. Apparently in *Naegleria*, the role of centrin is primarily for a differentiation- or flagellate-specific function. The temporary presence of centrin is concurrent with the presence of centriolar basal bodies, which supports the conjecture that in *Naegleria* centrin may be needed only when these organelles are present (Levy et al., 1998). Later studies performed full-genome transcriptional analysis at 20-min intervals throughout *Naegleria* differentiation and revealed vast transcriptional changes, including the differential expression of genes involved in metabolism, signaling and the stress response. Cluster analysis of the transcriptional profiles of predicted cytoskeletal genes revealed a set of 55 genes enriched in centriole components (induced early) and a set of 82 genes enriched in flagella proteins (induced late). The early set included genes encoding nearly every known conserved centriole component, as well as eight previously uncharacterized, highly conserved genes (Fritz-Laylin and Cande, 2010).

Notably, actin is not synthesized during the differentiation of amoebae into flagellates. Translatable actin mRNA begins to decrease in abundance within 7 min after the initiation of differentiation and thereafter decreases with a half-life of about 25 min. The selective disappearance of this major translatable mRNA provides a favorable opportunity to dissect the rules governing the half-life of a specific mRNA (Sussman et al., 1984b).

Amoebae, which lack microtubules except during mitosis, differentiate into flagellates with a fixed shape and a complex microtubular cytoskeleton within 120 min. The tension produced by an actomyosin network is required to maintain the flagellate shape. The rapid loss of the flagellate shape induced by drugs, which specifically block myosin light chain kinase, supports this hypothesis (Walsh, 2007).

The flagellar tubulin is synthesized *de novo* during differentiation (Fulton and Kowit, 1975). A novel alpha-tubulin gene (alpha6) is cloned. The open reading frame of alpha6 contained 1359 nucleotides encoding a protein of 452 amino acids (aa) with a calculated molecular weight of 50.5 kDa. The alpha6-tubulin is one of the most divergent alpha-tubulins so far known. Alpha6-tubulin is expressed in actively growing cells and repressed quickly when these cells are induced to differentiate. Immunostaining with an antibody against alpha6-tubulin showed that alpha6-tubulin is present in the nuclei and mitotic spindle-fibers but absent in flagellar axonemes or cytoskeletal microtubules. These data finally established the presence of an alpha-tubulin that is specifically utilized for spindle-fiber microtubules and distinct from the flagellar axonemal alpha-tubulins, hence confirmed the multi-tubulin hypothesis in this organism (Chung et al., 2002). The alpha-tubulin mRNA increases markedly in abundance during the first hour of differentiation, and then decreases even more rapidly with a half-life of approximately 8 min. The abundance of physical alpha-tubulin mRNA rises and subsequently falls in parallel with the abundance of translatable flagellar tubulin mRNA and with the *in vivo* rate of flagellar tubulin synthesis, which indicates that flagellar tubulin synthesis is directly regulated by the relative rates of transcription and mRNA degradation (Lai et al., 1988).

A gene that directs the programmed synthesis of flagellar beta-tubulin during the rapid differentiation from amoebae to flagellates is cloned. Beta-tubulin mRNA homologous to this gene family is expressed transiently during differentiation, and has not been detected in amoebae. The encoded beta-tubulin is strongly conserved, with features that closely resemble the beta-tubulins of diverse organisms, especially organisms that, like *Naegleria*, use tubulin to assemble flagellar axonemes. In most sequenced alpha-tubulins, the encoded carboxy-terminal amino acid is tyrosine, which undergoes post-translational removal and readdition, conserved processes of unknown function. In *Naegleria*, unusually, the terminus of alpha-tubulin is encoded as glutamine while that of beta-tubulin is tyrosine. The presence of these divergent termini on subunits of a conserved tubulin provoked us to re-examine aromatic amino acids at the termini of alpha- and beta-tubulins. This remarkable conservation of carboxy-terminal aromatic amino acids suggests that these residues serve some crucial function (Lai et al., 1994).

The distribution of two proteins, N-gammaTRP (*Naegleria* gamma-tubulin-related protein) and N-PRP (*Naegleria* pericentrin-related protein), was examined during the *de novo* formation of basal bodies and flagella that occurs during the differentiation. After the initiation of differentiation,

N-gammaTRP and N-PRP began to concentrate at the same site within cells. The percentage of cells with a concentrated region of N-gammaTRP and N-PRP was maximal (68%) at 40 min when the synthesis of tubulin had just started but no assembled microtubules were visible. When concentrated tubulin became visible (60 min), the region of concentrated N-gammaTRP and N-PRP was co-localized with the tubulin spot and then flagella began to elongate from the region of concentrated tubulin. When cells had elongated flagella, the concentrated N-gammaTRP and N-PRP were translocated to the opposite end of the flagellated cells and disappeared. The transient concentration of N-gammaTRP coincided with the transient formation of an F-actin spot at which N-gammaTRP and alpha-tubulin mRNA were co-localized. The concentration of N-gammaTRP and formation of the F-actin spot occurred without the formation of microtubules but were inhibited by cytochalasin D. These observations suggest that the regional concentration of N-gammaTRP and N-PRP is mediated by actin filaments and might provide a site of microtubule nucleation for the assembly of newly synthesized tubulins into basal bodies and flagella (Suh et al., 2002).

7.6. Differentiation-specific mRNAs

During the differentiation, four differentiation-specific mRNAs are transiently and coordinately accumulated. Three of the four mRNAs, Class II, III, and IV, encode alpha-tubulin, beta-tubulin, and flagellar calmodulin, respectively. Inhibition of protein synthesis at the beginning of differentiation completely blocked transcription of the beta-tubulin gene. Addition of cycloheximide at 30 or 40 min after initiation of differentiation inactivated transcription of the beta-tubulin gene in less than 10 min as judged by nuclear run-on experiments. However, once differentiation had proceeded for more than 50 min, inhibition of protein synthesis did not inactivate transcription of beta-tubulin mRNA was more active in cycloheximide-treated cells than in control cells. Cycloheximide treatment at the initiation of the differentiation also blocked transcription of the Class I gene. However, addition of the drug after 30 min had no significant effect on the transcription of the Class I gene. Cycloheximide treatment also increased the half-lives of beta-tubulin and Class I mRNA drastically. These findings suggested that: (i) the transient accumulation of the two differentiation-specific mRNAs during differentiation are regulated by changing both the rate of transcription and the stability of the mRNAs; (ii) protein synthesis is required for the transcriptional and post-transcriptional regulations; (iii) the transcriptional regulation mechanisms of the beta-tubulin gene and that of the Class I gene are distinct; and (iv) the transcription of the beta-tubulin gene is regulated by different mechanisms during differentiation (Bok et al., 1995). Three of the four mRNAs-alpha-tubulin, beta-tubulin, and Class I mRNA-began to be co-localized at the periphery of the cells as soon as transcription of the respective genes was activated and before any microtubular structures were observable. At 70 min after the initiation of differentiation, these mRNAs were re-localized to the base of the growing flagella, adjacent to the basal bodies and microtubule organizing center for the cytoskeletal microtubules. Within an additional 15 min, the mRNAs were translocated to the posterior of the flagellated cells, and by the end of differentiation (120 min), very low levels of the mRNAs were observed. Cytochalasin D inhibited stage-specific localization of the mRNAs, demonstrating that RNA localization was actin dependent. Since cytochalasin D also blocked differentiation, this raises the possibility that actin-dependent RNA movement is an essential process for differentiation (Han et al., 1997).

7.6.1. A calcineurin-B-encoding gene expressed during differentiation

Calcineurin B (CnB) is the regulatory subunit of calcium-calmodulin-regulated protein phosphatase 2B. CNB mRNA is readily detected in amoebae; its abundance increases fourfold during differentiation to flagellates, reaches a peak at

50-70 min, when flagella are forming, and then declines. A genomic clone matches an expressed cDNA, except that it is interrupted by two phase I introns. It is suggested that *Naegleria* is among the earliest branching eukaryotes known to contain canonical pre-mRNA introns (Remillard et al., 1995).

7.6.2. Two calmodulins in *Naegleria* flagellates

Two calmodulins have been described. Calmodulin-1, localized in flagella, has an apparent molecular weight of approximately 16 kDa, whereas calmodulin-2, localized in cell bodies, is 15.3 kDa. The finding of two calmodulins is unusual. Since the only known difference is apparent molecular weight, one calmodulin could be derived from the other, except that both calmodulins are synthesized in a wheat germ, cell-free system directed by RNA from differentiating *Naegleria*. Translatable mRNAs encoding calmodulins 1 and 2, not detected in amoebae, appear and subsequently disappear concurrently during the 100-min differentiation of *Naegleria* from amoebae to flagellates (Fulton et al., 1986). Both calmodulins are synthesized during differentiation from amoebae to flagellates; one remains in the cell body and the other becomes localized in the flagella. The single, intronless, expressed gene for flagellar calmodulin has been cloned and sequenced. The encoded protein is a typical calmodulin with four putative calcium-binding domains, but it has an amino-terminal extension of 10 divergent amino acids preceding conserved calmodulin residue 4. The transcripts encoding flagellar calmodulin and flagellate cell body calmodulin are clearly divergent. Expression of the flagellar calmodulin gene is differentiation-specific; its mRNA appears and then disappears concurrently with those encoding flagellar alpha- and beta-tubulin (Fulton et al., 1995).

7.6.3. CLP and CLB proteins

Two novel genes were cloned, Clp, Class I on plasma membrane and Clb, Class I at basal bodies, which are transiently expressed during differentiation and characterized their respective protein products. CLP (2,087 amino acids) and CLB (1,952 amino acids) have 82.9% identity in their amino acid sequences and are heavily *N*-glycosylated, leading to an ~ 100 × 10^3 increase in the relative molecular mass of the native proteins. In spite of these similarities, CLP and CLB were localized to distinct regions: CLP was present on the outer surface of the plasma membrane, whereas CLB was concentrated at a site where the basal bodies are assembled and remained associated with the basal bodies. Oryzalin, a microtubule toxin, inhibited the appearance of CLP on the plasma membrane, but had no effect on the concentration of CLB at its target site (Baek et al., 2012).

7.6.4. Nucleolar protein BN46/51

BN46/51 is an acidic protein found in the granular component of the nucleolus. When *Naegleria* amoebae differentiate into swimming flagellates, BN46/51 is found associated with the basal body complex at the base of the flagella. In order to determine the factors responsible for targeting BN46/51 to a specific subnucleolar region, cDNAs coding for both subunits are isolated and sequenced. Two clones, JG4.1 and JG12.1 representing the 46 kDa and 51 kDa subunits, respectively, were investigated. JG12.1 encoded a polypeptide of 263 amino acids with a predicted size of 30.1 kDa that co-migrated with the 51 kDa subunit of BN46/51 when expressed in yeast. JG4.1 encoded a polypeptide of 249 amino acids with a predicted size of 28.8 kDa that co-migrated with the 46 kDa subunit of BN46/51. JG4.1 was identical to JG12.1 except for the addition of an aspartic acid between positions 94 and 95 of the JG12.1 sequence and the absence of 45 amino acids beginning at position 113. Both subunits contained two KKE and three KKX repeats found in other nucleolar proteins and in some microtubule binding proteins. BN46/51 bound to polypeptides of 44 kDa and 74 kDa. The 44 kDa component was identified as the *Naegleria* homologue of fibrillarin. BN46/51 bound specifically to the nucleoli of fixed mammalian cells, which lack a BN46/51 related polypeptide. When the JG4.1 and JG12.1 cDNAs were expressed in yeast, each subunit was independently targeted to the yeast nucleolus. It was concluded that BN46/51 represents a unique nucleolar protein that can form specific complexes with fibrillarin and other nucleolar proteins. The association of BN46/51 with the MTOC of basal bodies may reflect its role in connecting the nucleolus with the MTOC activity for the mitotic spindle, which would provide a mechanism for nucleolar segregation during the closed mitosis (Trimbur et al., 1999).

7.6.5. NgUNC-119, *Naegleria* homologue of UNC-119, localizes to the flagellar rootlet.

The UNC-119 family of proteins is ubiquitous in animals. The expression of UNC-119 is prominent in neural tissues including photoreceptor cells. Homologues of UNC-119 are also found in ciliated (or flagellated) single-celled organisms. NgUNC-119 was not found in growing amoebae but accumulated rapidly after the initiation of differentiation into flagellates. Indirect immunofluorescence staining of differentiating *N. gruberi* showed that NgUNC-119 begins to concentrate at a spot near the nucleus of differentiating cells and then elongates into a filamentous structure. Purification and indirect immunofluorescence staining of the *Naegleria* flagellar rootlet suggested that NgUNC-119 is a component of the flagellar rootlet (Chung et al., 2007).

7.6.6. Thymidine kinase

Thymidine kinase activity decreases rapidly after the stimulation, while nucleoside phosphotransferase activity remains approximately constant (Bols et al., 1977). There is neither an activator of thymidine kinase in growing cells nor an inhibitor in differentiating cells. There is a strong correlation between the degree of differentiation and the degree of decrease in thymidine kinase activity. Addition of actinomycin D or cycloheximide does not prevent the decrease in thymidine kinase activity suggesting that neither transcription nor translation is involved in the putative inactivation of thymidine kinase enzyme. The residual thymidine kinase activity is primarily due to some portion of cells which did not respond to the stimulation and did not differentiate. Since there is an early termination of nuclear DNA synthesis in differentiating *Naegeria*, the decrease in thymidine activity (but no change in nucleoside phosphotransferase) suggests a correlation between thymidine kinase and DNA synthesis in *Naegleria* (Bols et al., 1977). When the growth temperature is shifted from 20°C to 32°C, a transient 2 - 4 fold increase of thymidine kinase activity occurs (Corff and Yuyama, 1978). These observations suggest that initiation of differentiation in *Naegleria* causes the cessation of thymidine kinase synthesis. This event is the earliest to be reported for the time course of *Naegleria* differentiation.

Later studies investigated the possible association of thymidine kinase with mitochondria by treating growing and differentiating cells with chloramphenicol, an inhibitor of mitochondrial protein synthesis (Corff et al., 1982). In some systems, chloramphenicol causes an overproduction of mitochondrial proteins coded for in the nucleus. It was observed that in growing *Naegleria*, chloramphenicol stimulates a dramatic increase in thymidine kinase activity while growth and division is gradually inhibited. Chloramphenicol does not stabilize the enzyme *in vivo* or *in vitro*. The stimulation is cycloheximide-sensitive and specific since nucleoside phosphotransferase activity does not increase. In cells stimulated to differentiate, chloramphenicol does not prevent differentiation or the expected decrease in thymidine kinase activity.

7.6.7. Heat shock

A heat shock to amoebae during flagellation results in the induction of heat shock proteins as well as multiple flagella (Walsh, 1980). The principal heat shock proteins are 96 kDa, 77 kDa, 70 kDa, and 68 kDa. These proteins are synthesized preferentially when cells at 25°C are shifted to temperatures above 32°C (Walsh, 1980). The maximal incorporation of methionine into heat shock proteins occurs at 38.2°C, the temperature at which maximal induction of multiple flagella has been reported (Walsh, 1980). Synthesis of heat shock proteins requires new transcription as judged by the ability of actinomycin D to inhibit their synthesis during the first 15 min of heat shock but not thereafter. Although heat shock can induce multiple flagella only when applied during a restricted interval, heat shock proteins are induced at any time cells are shifted to 38.2°C. The response to heat shock of the *Naegleria* heat shock proteins resembles that of *Drosophila* heat shock proteins, but the two groups of proteins differ in both size and number. *Naegleria* heat shock proteins are, however, strikingly similar in size to a group of heat-induced proteins found in chick embryo fibroblast, mouse L, and baby hamster kidney cells.

7.6.8. Effect of high hydrostatic pressure on transformation

The amoebae phase can be activated to transform to the flagellate phase, and cysts to excyst and transform to the flagellate phase, by a limited treatment with high hydrostatic pressure followed by release (Todd, 1972). The most effective treatment at 21 G is 45 min at 3500 psi (238 atm), which leads to almost 100% transformation. Following this dose of high pressure, 50% of amoebae transform within 55 - 70 min after release of pressure, and nearly all within 75 - 120 min. Nearly all cysts hatch and transform within 200 - 240 min after release. Pressures of 4000 psi (272 atm) and above, and of 1000 psi (68 atm) and below, were ineffective at any duration of treatment (Todd, 1972).

7.6.9. Effect of ions

The development of flagella by is a phenomenon related more to the ionic balance between the organism and its external environment than to the simple movement of water in and out of the cell in response to total osmotic changes. The presence of electrolytes inhibited the transformation of amoeba to flagellate, the molarity required varying with the salt used, namely 80 mM NaCl, 90 mM KCl, 50 mM $CaCl_2$ or 60 mM $MgCl_2$. Non-electrolytes also prevented this transformation at 250 mM for either sucrose or glucose, and this is known to be an osmotic effect. That the effect of ionic solutions was different was demonstrated by varying the time at which the environment was changed from distilled water to salt solution (Jeffery and Hawkins, 1975). Under higher concentrations, the organisms assume the amoeboid form. It is suggested that *Naegleria* assumes the amoeboid form when the external cation concentration exceeds the internal, and the flagellate form when the situation is reversed. The flagellate form may indeed be an adaptation towards cation conservation, while in the amoeboid form the cell is more concerned with preventing the entry of cations. Water movements in the opposite direction may, of course, also be involved, i.e. the flagellate form needs to eject water while the amoeboid form may need to conserve it. In both forms there is a contractile vacuole (Willmer, 1958).

7.6.10. Effect of bacterial suspensions

Bacterial suspensions inhibited transformation. Experiments with suspensions of either living or heat-killed bacteria in distilled water, together with the supernatants obtained when bacteria were removed by centrifugation, showed that the inhibition of transformation which occurred in bacterial suspensions was not due to any factors produced by the bacteria and present in solution. It appeared that this inhibition was brought about by the physical presence of the bacteria, either living or heat-killed (Jeffery and Hawkins, 1975).

7.6.11. Effect of β-mercaptoethanol

β-Mercaptoethanol added to *Naegleria* cultures at any time after the induction of flagellation, leveled the flagellation curve at a time 20 min later than the time of addition of the drug (Wade and Satir, 1968). When the β-mercaptoethanol was removed from blocked amoebae, there was subsequent recovery. Treatment with β-mercaptoethanol for 20 - 60 min caused excess delay in recovery such that the cells were setback to the time of induction. Comparison of the time of sensitivity to β-mercaptoethanol *versus* actinomycin D indicated that the primary effect of β-mercaptoethanol occurs at a later point in flagella morphogenesis than RNA synthesis. It was suggested that β-mercaptoethanol affects the assembly of precursor protein molecules into ordered structures and polymerization of microtubular or fibrous elements begins almost immediately upon induction of flagellation (Wade and Satir, 1968).

7.7. Flagellate to amoebae

Flagellates can "revert" to motile amoebae within 20 sec after a suitable stimulus, indicating that the amoeboid motility system remains latent in flagellates. A cell-produced chemical factor extracted from *Naegleria*, triggers a reproducible sequence of rapid shape changes in flagellates when added to their environment. Flagellates treated with 75% D_2O (which depolymerises microtubules) results in reversion to amoebae form within two min. This is associated with interruption of the more or less linear pattern of swimming in favour of rapid spinning, the onset of cytoplasmic streaming and consequent emission of pseudopods, which are able to make adhesion to the substrate (Preston and O'Dell, 1974). Electron microscopy of reverting cells revealed naked axonemes in the cytoplasm, the flagellar membrane possibly having fused with the cell membrane on penetration. This is interesting because fusion of the flagella with the cell does not occur during normal random contact and presumably reflected a recent membrane change accompanying reversion. By phase contrast microscopy these intracytoplasmic axonemes could be seen to continue to beat for a short while. Subsequently they could no longer be resolved, the integrity of their microtubules having broken down (Preston and O'Dell 1974).

8

Growth and Life Cycle

Abstract

N. fowleri belongs to the genus *Naegleria*, family Vahlkampfiidae in the class Heterolobosea. It is a free-living protist that feeds mostly on bacteria. *N. fowleri* has a simple life cycle that includes asexual reproduction by binary fission; however its ability to respond to changing environments and remain viable is both complex and intriguing. While there are invasive free living stages, proliferation and differentiation are highly regulated events. Here, we describe the present knowledge of aspects of the pathogen's life cycle, and factors associated with influencing its growth.

8.1. Food selection and ingestion

Naegleria feed on yeast, algae and both Gram-negative and Gram-positive bacteria (Anderson and Jamieson, 1974). Xinyao et al., (2006) showed that filamentous cyanobacteria (e.g., *Anabaena*, *Cylindrospermum*, *Gloeotrichia*, and *Phormidium*) are readily consumed. However, the tight threads (*Oscilltoria*) and aggregates (*Aphanizomeno*n) are not ingested; however, their sonicated fragments are observed inside food vacuoles, suggesting that their morphologies prevent them from being ingested. Notably, unicellular Chroococcaceae (e.g., *Synechococcus*, *Aphanocapsa*, and *Microcystis*) are excreted after ingestion, indicating that food selection takes place inside food vacuoles. Ingestion strongly depended on the satiation status of the amoebae, starved amoebae fed at higher rates compared with satiated amoebae (Xinyao et al., 2006).

8.2. Cultivation of *N. fowleri*

N. fowleri can be grown simply on the surface of non-nutrient agar (1.5%) spread with living or dead *Enterobacter aerogenes* or *E. coli* or other Gram-negative bacilli (Anderson and Jamieson, 1974). Live bacteria support optimal growth compared with heat-killed bacteria (Anderson and Jamieson, 1974). Under these conditions, the ameobae feed upon the bacteria, and as growth enters stationary phase and the food supply is used up, they begin to encyst. Cysts, if kept from drying out, will remain viable for months, possibly years. In contrast to *N. gruberi*, *N. fowleri* grows best in less enriched media. Cerva medium contains 2% Bacto-Casitone (Difco) and 10% fresh horse serum in distilled water (Cerva, 1969) and Nelson medium has liver digest, glucose, and calf serum. Cerva medium gives optimal growth in unagitated cultures (Haight and John, 1980). By using agitated cultures and Nelson medium, it is possible to obtain large quantities (3 x 10^9 amoebae per liter) of *N. fowleri*. At 37°C, the mean generation time is 5.5 h for exponentially growing cells. There is only slight utilization of glucose, and amino acids appear to serve as carbon and energy sources. The pH optimum for growth initiation in agitated cultures is 5.5, and 6.5 in unagitated cultures (reviewed in John, 1982). Serum appears to be an important component of the liquid media used for axenic cultivation of *N. fowleri*. Of the 17 sera tested, heat-inactivated calf serum supported the greatest cell yields, whereas heat-inactivated fetal calf serum produced the lowest cell yields. Between these two and ranked in order of decreasing cell yield are pig, dialyzed calf, monkey, newborn calf, lamb, turtle, dog, chicken, mouse, rabbit, frog, horse, gamma globulin-free calf, fish, and human sera. The recommended growth medium is Nelson's medium (0.4 mg of $MgSO_4$, 0.4 mg $CaCl_2$, 14.2 mg Na_2HPO_4, 13.6 mg KH_2PO_4, 12 mg NaCl, in 100 mL of distilled water and then addition of 0.17 g liver infusion, 0.17 g glucose. This should be autoclaved for 25 min at 121°C, followed by the addition of sterile heat inactivated foetal calf serum (final concentration of 10% serum) just prior to use.

8.3. Chemically defined medium

Chemically defined minimal medium for the cultivation of high temperature tolerant and pathogenic *Naegleria* spp. have been developed. A defined minimal medium for *N. fowleri* consists of eleven amino acids (arginine, glycine, histidine, isoleucine, leucine, methionine, phenylalanine, proline, threonine, tryptophan, and valine), six vitamins (biotin, folic acid, hemin, pyridoxal, riboflavin, and thiamine), guanosine, glucose, salts, and metals. The addition of protoporphyrin IX to Nelson medium resulted in a modest increase in mobility, resistance to complement lysis and virulence when compared to *N. fowleri* amoebae grown in Nelson medium without added porphyrin (Nerad et al., 1983; Bradley et al., 1996).

8.4. Cell density within the biofilm

The cell density of *N. fowleri* within biofilms is significantly affected both by the temperature and the nutrient level (bacteria:amoeba ratio). Goudot et al., (2012) showed that at 32°C, the density remained constantly low (1-10 *N. fowleri* per cm^2) indicating that the amoebae were in a survival state, whereas at 42°C the density reached 30-900 *N. fowleri* per cm^2 indicating an active growth phase. A threshold for the nutrient level of close to 10^4 bacteria/amoeba is needed to detect the growth of *N. fowleri* in freshwater biofilm (Goudot et al., 2012).

8.5. Effect of pH, viscosity on *N. fowleri* growth

The pH optimum is 6.5 for *N. fowleri* and 6.0 - 6.5 for *N. gruberi* (Cerva, 1978). No negative influence on the growth of *N. fowleri* is observed even at 0.5% concentration of highly viscous methylcellulose, whereas the growth of *N. gruberi* is distinctly inhibited by more than 0.2% of methycellulose. The growth of *Naegleria* species was inhibited by 0.1 N NaCl and KCl. The inhibitory effect of these salts correlated primarily with the concentration of chloride anion.

8.6. Effect of porphyrin on *N. fowleri* growth

The iron-containing porphyrins, hemin or hematin, or the iron-free porphyrin, protoporphyrin IX, are effective in supporting growth of *N. fowleri* in Cline medium lacking serum. Iron-binding proteins, including hemoglobin, do not satisfy the growth requirement of the amoebae for exogenous porphyrin. Expression of biological functions including azocaseinase activity, agglutination, mobility, complement susceptibility, and virulence are altered by the composition of the growth medium. Amoebae grown in Cline medium supplemented with either hemin or protoporphyrin IX display greater mobility and are more resistant to lysis by Complement than those grown in Nelson's medium. Similarly, amoebae grown in Cline medium supplemented with either hemin or protoporphyrin IX are more pathogenic for B6C3F1 mice than those grown in Nelson medium.

8.7. Cell cycle

The amoeboid form reproduces by binary fission and also gives rise to the cyst and flagellate form. Neither cyst form nor flagellate form reproduces. It was established from cell cycle analysis that the approximate intervals of G1, S, G2, and M phases were: 6% M phase (28 min), 38% G1 phase (180 min), 38% S phase (183 min), 19% G2 phase (90 min), with a total cell cycle time of 8 h. However, shorter estimates of S phase are obtained for monoxenic cultures (Corff and Yuyama, 1976).

Throughout the stages of mitosis, the nucleolus is present. During metaphase, several deeply stained DNA condensations following an elongated pattern are observed, corresponding almost certainly to tightly grouped chromosomes. Ultrastructural observations demonstrated that the nucleus divides by cryptomitosis, a process in which the nuclear membrane does not disappear during the mitosis. Centrioles are not found, and a spindle of microtubules is observed running the length of the nucleus from pole to pole however, they do not come to a focal point (González-Robles et al., 2009).

8.8. Respiration in *N. fowleri*

N. fowleri infects an oxygen-rich environment (the brain) and so one would expect it to have an aerobic metabolism. *N. fowleri* lives in aerobic aqueous environments and has many mitochondria. Whole cell respiration rates were measured polarographically throughout the growth cycle of *N. fowleri*. Under agitated culture conditions, amoebae consumed 30 ng of O_2 per min per mg of cell protein during log growth. Under similar conditions *N. gruberi* amoebae consumed 80 ng of O_2 per min per mg of cell protein. The lower oxygen consumption, and most likely oxygen requirement, by *N. fowleri* probably explains the presence of the pathogen in heated waters where dissolved oxygen concentrations are substantially reduced. The respiratory rate gradually declined

(3-fold) during stationary growth phase. The reduction in respiratory rate may involve respiratory control since further increases in respiratory rate does not occur in spite of fresh oxygen supplies. The respiratory process of isolated *N. fowleri* mitochondria is similar to classical mammalian cell mitochondria (John, 1982). Mitochondria rapidly oxidized glutamate, NADH, pyruvate, succinate, and other Krebs cycle intermediates but slowly oxidized lactate and a-glycerophosphate. The rates of substrate oxidation are ADP dependent and phosphorylative efficiencies (ADP:O ratios) are about 1.4 for NAD-linked substrates and 1.0 for succinate. The respiratory control ratios were 1.5 to 3 for 11 substrates and dependent on the addition of Pi, Mg^{2+}, and serum albumin to the reaction mixture. Cyanide, azide, malonate, and amytal inhibited the oxidative phosphorylation of the present mitochondria essentially the same way as that of the mammalian system while rotenone inhibited both glutamate and succinate oxidation. Pentachlorophenol and DNP uncoupled glutamate and succinate oxidation from phosphorylation. Difference spectra of oxidized and dithionite-reduced mitochondria show distinct absorption bands of flavins, *c*-type, *b*-type, and *a*-type cytochromes (Weik and John, 1979). Whole cell respiration rates are measured polarographically during agitated cultivation of *N. fowleri* (Weik and John, 1979). During log growth, amoebae consumed 30 ng atoms O per min per mg cell protein. The amoebae respiration rate gradually decreased 3-fold during stationary phase. Intact mitochondria are isolated from *N. flowleri* and the oxidative and phosphorylative capacities of the mitochondria are studied with an oxygen electrode apparatus. Mitochondria, stored in isolation medium at 0°C, lost about 20% of respiratory and phosphorylating activity after 4 h and were most stable at pH 7 to 8 (Weik and John, 1979).

Notably, the genome of *N. gruberi* demonstrated that it is extremely versatile, and although considered to be fully aerobic, its genome also predicts anaerobic respiration (Ginger et al., 2010; Opperdoes et al., 2011; Ginger et al., 2010; Tsaousis et al., 2014). *N. gruberi* has a functional [FeFe]-hydrogenase, an enzyme that is attributed to anaerobic organisms. Surprisingly, hydrogenase is localized exclusively in the cytosol, while no hydrogenase activity was associated with mitochondria of the organism. In addition, cytosolic localization displayed for HydE, a marker component of hydrogenase maturases. *Naegleria*, an obligate aerobic organism and one of the earliest eukaryotes, is producing hydrogen, a function that raises questions on the purpose of this pathway for the lifestyle of the organism (Tsaousis et al., 2014). The carbohydrates are oxidised to carbon dioxide and water when oxygen is not limiting and that in the absence of oxygen the end-products will be succinate, acetate and minor quantities of ethanol and D-lactate. The hybrid mitochondrion/hydrogenosome has both cytochromes and an [Fe] hydrogenase, but seems to lack pyruvate-ferredoxin oxidoreductase. These studies suggest that *N. gruberi* is well-equipped for ATP production using a facultative anaerobic metabolism.

8.9. Storage

8.9.1. Cryopreservation

The cryopreservation of *Naegleria* from axenic or monoxenic culture with *E. coli* can be achieved using a two-step cooling process in freezers and storage in liquid nitrogen. Trophozoites can be suspended in axenic culture medium and dimethylsulphoxide (DMSO) added to a final concentration of 5%. Ampules are placed directly at -20°C for 60 min, followed by a further 60 min at -70°C and then stored in liquid nitrogen. On rapid thawing of the ampules at 37°C, recovery rates are approximately 8% for *N. fowleri* (Kilvington and White, 1991).

Later, John et al., (1994) tested a variety of conditions of cryopreservation. The average best conditions for freezing was 1 x 10^6 exponentially growing amoebae per mL of freezing medium consisting of 12% DMSO, 20% heat-inactivated bovine calf serum, 4% glucose; 30 min equilibration at 23°C (room temperature), followed by 60 min at -20°C, with storage at -70°C. Under these conditions viability after 1 month of freezing was 64% for *N. fowleri*. After 12 months of freezing, viability was 47% for *N. fowleri* (John et al., 1994). At 5 years of cryostorage viability was 38% for *N. fowleri*. The virulence of *N. fowleri* did not decrease during 30 months of freezing (John and John 1996). At 10 years of cryostorage, viability was 21% for *N. fowleri* (John and John, 2006).

9

Ecology

Abstract

The knowledge of the ecology of *N. fowleri*, its environmental niches, the effect of climate change or thermal pollution, or the role of man-made environments such as recreational waters, on the growth and distribution of amoebae abundance in the environment is critical to our understanding of its biology, as well as to develop preventative measures and create public awareness. To this end, *N. fowleri* is ubiquitous. It has been isolated from diverse natural environments including soil, water, and air, as well as man-made environments, and human tissues. Here, we summarize the distribution of *N. fowleri* in various environments.

9.1. Free-living amoebae

Being a free-living amoeba, *N. fowleri* is present in diverse environments including soil, water, and air. *N. fowleri* have been detected on all continents, except Antarctica (De Jonckheere, 2011). However, other species of *Naegleria* have been isolated from Antarctica. Samples of soil and water were taken from the McMurdo Sound-Dry Valley region of the Antarctica. Although *Acanthamoeba* isolates appeared to show better survival potential, *Naegleria* were isolated (Brown et al., 1982). Later De Jonckheere (2006) isolated novel species of *Naegleria* and *N. antarctica*, *N. dobsoni* and *N. chilensis* are their closest relatives from freshwater samples with sediment taken from two regions in the Arctic, Spitzbergen and Greenland, and one region in sub-Antarctica, Ile de la Possession. It was suggested that the *Naegleria* gene pool present in the polar regions is different from that found in temperate and tropical regions (De Jonckheere, 2006).

9.2. Isolation from the atmosphere

Using an air vacuum sampler and filters, the presence of free-living amoebae was determined (De Jonckheere, 1977). The species isolated included *Naegleria* spp. *Acanthamoeba* spp. *Vahlkampfia* spp. *Saccamoeba* spp. *Hyalodiscus* spp., *Platyamoeba* spp., *Rugipes* spp., *Vannella* spp., and *Leptomyxa* spp. Among the species isolated *Naegleria* spp. and *Acanthamoeba* spp. include strains, known to be pathogenic in humans (Rivera et al., 1987). Later studies recovered soil amoebae from the air during the harmattan in Zaria, Nigeria. Non-nutrient agar plates seeded with *E. coli* were used as settle plates and exposed to the air for 30 min to four hours, after which they were incubated at 27°C or at 37°C (Lawande, 1983). A total of 38 strains of amoebae were cultured: 21 of the genus *Hartmannella*, eight *Naegleria*, four *Schizopyrenus*, three *Didascalus* and two *Tetramitus* (De Jonckheere et al., 1997). Pathogenic *N. fowleri* were recovered that killed mice. These results suggest the possibility of airborne primary amoebic meningoencephalitis infections. They have also been isolated from water-hyacinth root (*Eichhornia crassipes*) (Ramirez et al., 2010).

9.3. Isolation from freshwater lakes

N. fowleri is widespread in freshwater lakes (Wellings et al., 1977). Populations in three of five lakes sampled routinely reached levels of one amoeba per 25 mL of water tested during the hot summer months (Wellings et al., 1977).

When tested for seasonal distribution, population densities of free-living amoebae peaked in late summer. Littoral sediment was the major habitat for free-living amoebae, with peaks in populations of *Acanthamoeba* and *Naegleria* in August, *Hartmannella* in July, and *Vahlkampfia* in May. Distribution patterns and species composition of free-living amoebae from surface water were similar to those from littoral sediment; however, a greater percentage of *Naegleria* was found in surface water. Numerous free-living amoebae were isolated from the neustonic community (surface film), and the number of free-living amoebae isolated in the surface film at the deep water station was higher than the number from subsurface (5-10 cm) samples. Free-living amoebae populations consistently were highest in the detrital layer, which persisted at a depth of 3.0-3.4 m throughout the summer period. The large percentage of *Naegleria* contributing to free-living amoebae in the detrital layer suggests that *Naegleria* sink through the layer, flagellate, and swim back up, such migrations possibly being triggered by a reduction of nutrients below the layer or by the presence of anoxic, reducing conditions in the hypolimnion. In addition, weather events were found to play a major role in the redistribution of free-living amoebae between various habitats in the aquatic ecosystem, with such changes probably due to resuspension of free-living amoebae from littoral sediment by wind action and input from the watershed via runoff (Kyle and Noblet 1986). *Naegleria* and *Vahlkampfia* were the most frequently encountered free-living amoebae in littoral sediment and surface water samples whereas *Acanthamoeba* was most commonly isolated from profundal sediment, especially during late summer. In the water column, free-living amoebae populations were highest in a persistent detrital layer; however, few amoebae were isolated from a massive (approximately 1.5 m thick) layer of Oscillatoria. *N. fowleri* was isolated from the detrital layer (Kyle and Noblet, 1987).

N. fowleri were isolated from river water, albeit, the recovery was better for cysts than for trophozoites (Pernin et al., 1998).

9.4. Prevalence of *Naegleria* and wild animals

Thirteen wild mammalian species were tested and shown to be serologically positive for anti-*Naegleria* spp. antibodies with varying degrees of titers indicating that wild mammals have differing degrees of contact with *Naegleria* spp. based on ecological or behavioral adaptations. It was shown that if a mammal is exposed to *Naegleria* spp. in the environment, titers of anti-*Naegleria* spp. antibodies will not significantly differ, regardless of age or sex. Adults of many species had significantly higher occurrences of anti-*Naegleria* spp. antibodies. Populations of juvenile wild mammals are probably at higher risk than adults to *Naegleria* infection in the environment, not because of lower titers, but because their chance of having anti-*Naegleria* spp. antibodies is less than adults (Kollars and Wilhelm, 1996).

In other studies, a total of 508 reptiles were captured at Canary Islands (Spain) and examined for free-living amoebas. Two hundred seventy-three clones of amoebas were isolated by culture of gut contents, 157 of them belonging to the genus *Acanthamoeba* and 12 to the genus *Naegleria* (Sesma and Ramos, 1989). Pathogenic *Naegleria* spp. have also been isolated from organs of freshwater fish (Taylor, 1977; Dyková et al., 2001).

9.5. Distribution of *Naegleria* from clinical samples and clinical settings

The moist areas in physiotherapeutic departments of hospitals were investigated for the occurrence of *Naegleria* and *Acanthamoeba*. Notably, 61% of the swabs taken were positive with one or several species of amoebae (Michel and Menn, 1991). *Naegleria* and *Acanthamoeba* were isolated from dust samples in hospitals in Brazil. Hospital collection sites included intensive care unit, operation rooms, nursery, kitchen, emergency and infectious diseases isolation room. *Acanthamoeba* and *Naegleria* were found in 45.5% of the samples, of which 41.6% were collected in the university hospital and 50% in the state hospital. Of all, 45.5% were positive for the genera *Acanthamoeba* and 3.8% for genera *Naegleria* (da Silva and da Rosa 2003).

Among dental treatment units tested, the incidence of one or more amoeba species was observed in all water-carrying systems of the dental treatment units. In 8.2% of the units, *Naegleria* species was found and in 12.2% *Acanthamoeba* species was present (Michel and Just, 1984). All the dental unit water samples tested contained amoebae at concentrations up to 330 per mL, or more than 300 times the concentration in tap water from the same source. *Hartmanella*, *Vanella*, and *Vahlkampfia* spp. were the most frequently encountered. *Naegleria* and *Acanthamoeba* spp. were also present in 40% of the samples. It was suggested that biofilms that form inside some dental instruments can considerably increase the concentration of free-living amoebae (Barbeau and Buhler 2001).

9.6. Nasopharyngeal and oral regions of dental patients

N. fowleri have been isolated from swabs taken from nose, mouth, and pharynx, suggesting that healthy patients are carriers of pathogenic protozoa (Rivera et al., 1984). Following a fatal case of primary amoebic meningo-encephalitis during the dusty harmattan period in an 8-month-old child, a survey was carried out to examine the nasal passages of children for the presence of soil amoebae during the harmattan. In all, 50 children were evaluated for the presence of soil amoebae. Positive cultures for the soil amoebae were obtained from 12 children (24%). Four species of amoebae were isolated singly or in combination with other species. *Naegleria fowleri* that proved pathogenic for mice, were cultured from specimens from two children. Later studies tested local populations of Zaria, to find out the incidence of free-living amoebae in their nasal passages (Abraham and Lawande, 1982). The times of sampling were spaced so as to cover both the rainy (non-harmattan) and dry (harmattan) seasons. In all, 1,250 individuals were sampled, and were grouped in the three age groups of above 18 years, between 4 and 18 years and below 4 years. The overall incidence was 4.2% (52 out of 1250). There was no marked difference in the three age groups studied. The incidence rate in males was 4.8% (30 of 630) and that in females was 3.5% (22 of 620). Nine different species of free-living amoebae were isolated. Six belonged to the genus *Hartmannella*, two to the genus *Naegleria*, and one to the genus *Schizopyrenus*. *N. fowleri* were pathogenic for mice. It was observed from this study that a significant percentage of the Zaria population carry free-living amoebae in the nasal passages. The monthly incidence rate in population ranged from 1.8 to 3.1% during the rainy (non-harmattan) season whereas in the dry (harmattan) season it ranged from 4.2 to 7.9%. The highest incidence rate coincided with the peak of the dry (harmattan) season.

9.7. Serology of *Naegleria* spp.

Antibodies to *N. fowleri* have been detected in surveys of normal human sera and saliva with titres ranging from 1:5 to 1:20 (Cursons et al., 1980b; reviewed in John, 1982). The antibodies belonged to IgA, IgG, and IgM. These antibodies could participate in the resistance to the infection, probably by inhibiting the adherence of *N. fowleri* trophozoites to the nasal mucosa (Cursons et al., 1980b; reviewed in John, 1982; Rivera-Aguilar et al., 2000; Rivera et al., 2001).

In one study, antibodies were found in all 200 sera tested from New Zealand with titres ranging from 1:5 to 1:20 (Cursons et al., 1980). In another study, antibodies were detected in 101 specimens from 115 hospital patients ranging in age from 15 to 98 years. Class-specific anti-immunoglobulins identified antibodies as IgG and IgM. IgG antibody titers to both species ranged from 1:20 to 1:640. Seven of 15 serum samples collected from newborn infants also demonstrated IgG antibodies to these organisms with a titer range of 1:20 to 1:80. The immunoperoxidase test and Western blot analysis of selected serum samples demonstrated a close similarity in serological results between *N. fowleri* and *N. lovaniensis* (Dubray et al., 1987). Later studies tested serum samples of 1,054 inhabitants of Bohemia (Czechoslovakia) and showed the frequency of positive reactions of up to 3.5 per cent (Cerva, 1989).

9.8. Effect of thermal pollution on the distribution of *N. fowleri*

The occurrence of *N. fowleri* is often associated with elevated temperatures and/or industrial wastewater. *N. fowleri* showed 100% survival at pH ranging from 2.1 to 8.15 (Sykora et al., 1983). The discharges of 16 thermal polluting factories were examined for the occurrence of *Naegleria fowleri* (De Jonckheere and Voorde, 1977) and shown to harbor this amoeba. *N. fowleri* isolates were highly virulent for mice when inoculated intranasally. More *N. fowleri* were found during summer than in winter. The distribution seems to be bound to the cooling waters of older factories and is not restricted to one type of factory (De Jonckheere and Voorde, 1977). Artificial heating of water by power plant discharges facilitates proliferation of *N. fowleri* (Dive et al., 1981), however *N. fowleri* was not detected outside the reach of the thermal pollution (Cerva et al., 1982), suggesting that warm discharge water should under no conditions be used directly for sports and recreational purposes.

Kasprzak et al., (1982) tested two complexes of lakes and canals supplying water for two electric power plants, their steam condensors and an adjoining river for the presence of *N. fowleri*. Sixty-four strains of *N. fowleri* were isolated, 13 isolates being virulent for mice when instilled intranasally. These strains were found in the steam condensor of the power station A and in waters polluted with warm water of this plant. *N. fowleri* strains occurred also in an adjoining river connected with the water system of the power plant. The results show the possible role of the steam condensor A as an incubator and regular source of pollution with pathogenic amoebae for its own system of cooling waters and even the adjoining river (Kasprzak et al., 1982). During periods of thermal additions to cooling towers, the concentration of *N. fowleri* increased as much as 2 orders of magnitude. Concentrations of amoebae returned to prior thermal perturbation levels within 30 to 60 days after cessation of thermal additions (Tyndall et al., 1989).

A canal draining cooling water from a factory showed presence of *N. fowleri* in scrapings off the canal walls and in its bottom sediment for a length of about 2 km starting at the site of the outlet of the water from the factory. The maximum number of amoebae in one liter of the sample was 800 individuals (Cerva et al., 1980). *N. fowleri* has been isolated from sewage samples (Bose et al., 1990). Now it is widely accepted that polluted environments are the principal sources of potentially pathogenic species of free-living amoebae. This was explained by Griffin (1983), flagellate-empty habitat hypothesis, i.e., human intervention and/or natural events remove usual competitors and the ability to transform to a motile flagellate confers an advantage to *N. fowleri* in recolonizing.

9.9. *N. fowleri* isolation from swimming pools

Swimming pool waters of Mexicon city showed the prevalence at 16.77% pools tested. *Naegleria* were recovered in their cyst forms. Swimming pools of the indoor type with an inner side garden require higher free-chloride residue. In such pools, the free residual chloride values of 0.50 to 1.5 mg/liter, ordinarily used in pool waters, may not be adequate for elimination of amoebae cysts (Rivera et al., 1983), even though a free chlorine residual of 1.0 mg per liter and a pH range of 7.0 to 7.6 are recommended by local health authorities and often effective (Esterman et al., 1984). However, free-chloride levels of 5.31 mg per mL or more is shown to be highly effective (Rivera et al., 1993). A high level of chloride would counter additional factors such as soil contamination. For example, Kadlec et al., (1978) located a reservoir of *N. fowleri* in the cracked wall of a swimming pool where repeated outbreaks of primary amoebic meningo-encephalitis were observed between 1962 and 1965.

9.10. *Naegleria fowleri* in the thermal recreational waters

Naegleria fowleri is often found in most geothermal baths or in the proximity of the baths. An epidemiological survey of thermal baths and mud-basins showed that over 51 samples of water and mud incubated at 37°C and at 45°C, 34 (66.7%) became positive at 37°C and 33 (64.7%) at 45°C. Among different amoebae, 7 (6%) strains of *Naegleria* spp., 6 (5.2%) of *Acanthamoeba* spp., 39 (33.6%) of *Vahlkampfia* spp., 28 (24.1%) of *Hartmannella* spp. and 36 (31.1%) strains of other species of free-living amoebae were isolated. Four strains of *Naegleria* spp. and six of *Acanthamoeba* spp. proved pathogenic in animals (Scaglia et al., 1987).

The contamination of the water with *N. fowleri* likely occurs after emerging from the geothermal source, when the water runs over the soil. Therefore, it should be possible to reduce the concentration of *N. fowleri* in the geothermal baths by installing a pipeline between the geothermal sources and the baths and by preventing flooding water from entering the baths after rainfall. This measure were proven effective in eliminating *N. fowleri* from a pool located inside a re-education clinic (Moussa et al., 2014). Additionally, *N. fowleri* has been isolated from hot springs in Yellowstone and Grand Teton National Parks (Sheehan et al., 2003).

9.11. Assays for the identification of *N. fowleri* in environmental water samples

Identifying sites contaminated by *N. fowleri* is important in order to prevent the disease. Although culture-based assays are reliable, a major limitation of culture-based assays is that it detects both nonpathogenic and pathogenic species, and does not account for *N. fowleri* cells. Reveiller et al., (2003) developed ELISA for the specific identification of *N. fowleri* in primary cultures of environmental water samples. Of the 939 samples isolated from artificially heated river water and screened by ELISA, 283 were positive, suggesting that ELISA method is reliable and can be considered as a powerful tool for the detection of *N. fowleri* in environmental water samples (Reveiller et al., 2003).

Flow cytometry has proven efficient for identification of *N. fowleri* trophozoites and cysts from river and surface water samples (Muldrow et al., 1982). With this system, it is possible to rapidly identify species and quantitate mixtures of pathogenic amoebae in environmental samples. Cytofluorographic analysis of amoebic isolates reduces the time presently required to screen environmental sites for pathogenic amoebae. Flow cytometry is advantageous over fluorescent microscopy in that it includes a high degree of statistical precision due to the large numbers measured, high immunofluorescent titers, and elimination of subjectivity and fluorescence fading (Muldrow et al., 1982; Pougnard et al., 2002; Johnson et al., 2007).

A species-specific PCR for the identification of *N. fowleri* has been developed. In sensitivity studies, 10 trophozoites or cysts and one trophozoite or cyst could be detected after 35 and 45 cycles, respectively. In conjunction with a rapid DNA isolation method, this PCR was used to identify *N. fowleri* directly from primary cultures of environmental samples (Kilvington and Beeching, 1995). PCR-based assays have been developed successfully to identify *N. fowleri* from water samples (Marciano-Cabral et al., 2003; Maclean et al., 2004). Sparagano (1993b) used two assays to extract DNA from samples: first, direct DNA extraction, which gave positive results only for water samples without sediment; second, DNA extraction after sample incubation on agar plates, which gave positive results for all samples. Thus, this molecular identification is a powerful tool to investigate *N. fowleri* growth in environmental samples (Sparagano, 1993b). A multiplex PCR has also been developed to simultaneously detect *N. fowleri* and other *Naegleria* species in the environment. Multiplex PCR was also capable of identifying *N. fowleri* isolates with internal transcribed spacers of different sizes (Pélandakis and Pernin, 2002).

Recently, a duplex real-time PCR (qPCR) method to detect and quantify *N. fowleri* in water samples has been developed. In this assay, primers were designed based on the Mp2Cl5 gene to quantify *N. fowleri* DNA in a single duplex reaction. The qPCR detection limit (DL) corresponds to the minimum DNA quantity showing significant fluorescence with at least 90% of the positive controls in a duplex reaction. Using fluorescent Taqman technology the qPCR was found to be 100% specific for *N. fowleri* with a DL of 3 *N. fowleri* cell equivalents and a PCR efficiency of 99%. Although qPCR is well suited for rapid and quantitative detection of *N. fowleri* in real water samples, nevertheless "low contamination levels" of water samples (<200 *N. fowleri* cells per litre still require culture method analyses (Behets et al., 2007). Puzon et al., (2009) tested a total DNA extraction technique coupled with a real-time PCR method using primers specific for *N. fowleri* and readily detected *N. fowleri* without preculturing with the lowest detection limit for *N. fowleri* cells ranging from one to 10 cells (Puzon et al., 2009).

Ahmad et al., (2011) developed a nested one-step PCR test, in conjunction with a direct DNA extraction from water or sediment material, for the rapid and reliable detection of *N. fowleri* from the environment. Mull et al., (2013) used an immunomagnetic separation (IMS) procedure and real-time PCR TaqMan assay to recover and quantify *N. fowleri* in water and sediment samples. When one liter of lake water was seeded with *N. fowleri*, the method had an average recovery of 46% and detection limit of 14 amoebas per liter of water. The method was then applied to sediment and water samples with unknown *N. fowleri* concentrations, resulting in positive direct detections by real-time PCR in 3 out of 16 samples and confirmation of *N. fowleri* culture in 6 of 16 samples (Mull et al., 2013). Kao et al., (2014) used PCR of 5.8S rRNA gene and ITS region to successfully identify *Naegleria* isolates and quantify the *Naegleria* spp. by TaqMan real-time quantitative PCR in reservoir water samples. Overall, these rapid and reliable detection tools should prove useful to facilitate studies of the physical, chemical, and biological factors associated with the presence and dynamics of *N. fowleri* in environmental systems.

10

War of the Microbial Worlds

Abstract

N. fowleri harbours microbial organisms including viruses, and bacteria, some of which are potential pathogens. The precise nature of this symbiosis is not clear, but it is suggested that such interactions enable pathogenic microbes to survive hostile conditions and this association can lead to their transmission to susceptible hosts to establish infection. In this context, *Naegleria* can act as vehicles for the multiplication and dispersal of pathogenic microbes as well as providing shelter for such bacteria from antibiotic and disinfection treatments. These findings suggest that employing anti-amoebic approaches in eradicating bacterial pathogens may be an effective strategy to counter threat from both *Naegleria* as well as bacterial pathogens. For example, *Naegleria-Legionella* interactions have gained significant attention by the scientific and the medical community.

10.1. A host for virus-like particles

Schuster (1969) observed intranuclear virus-like bodies in cultures of *Naegleria* from soil. The particles were ~100 nm in diameter, and mainly restricted to the nucleus of the cell. Passage of particles from the nucleoplasm into the cytoplasm was suggested by their association with tubular projections from the nuclear membrane, and particles were seen in the cytoplasm of the amoebae. It was suggested that the virus-like bodies resembled reovirus. *Naegleria* containing virus-like particles exhibited large bundles of microtubule-like fibrils in the nucleoplasm, and, in addition, the nuclei showed degenerative modifications (Schuster and Dunnebacke, 1974), while amoeba cytoplasm exhibited regions of varying densities and bundles of microtubule-like fibrils (Schuster and Dunnebacke, 1974b). The subcellular infectious material in *Naegleria* was capable of infecting chick embryo cells and causing them to undergo cytopathic changes with the release of more infectious material. The cytopathogenic agent was filterable and passageable and present in the strains of the *Naegleria* whether they are obtained free-living from soil samples (*N. gruberi*) or as pathogens from humans (*N. fowleri*). The agents obtained from the different *Naegleria* strains are similar in amount and in their cytopathogenic interaction with chick cultures (Dunnebacke and Schuster, 1971; Dunnebacke and Schuster, 1974; Schuster and Clemente, 1977).

10.2. *N. fowleri* and bacteria interactions

In addition to producing fatal infections, *N. fowleri* are recognized as environmental hosts, indirectly involved in the epidemiology of many pathogenic bacteria. Phillips (1974) reported the presence of intracellular diphtheroids. The organisms could be eliminated by antibiotic treatment but amoebae grew better in its presence suggesting a symbiotic relationship. Studies of the interaction of *Legionella pneumophila* with free-living amoebae showed that *Naegleria* could use *L. pneumophila* as a sole food source (Tyndall and

Domingue, 1982). However, growth of the amoebae on non-nutrient agar plates seeded with *L. pneumophila* was slower than growth on nonnutrient agar plates seeded with *Escherichia coli*. On inoculation of *L. pneumophila* into axenic cultures of *Naegleria*, 99.9% of the *L. pneumophila* were destroyed within 24 h. After several weeks, however, some amoeba cultures became chronically infected and supported the growth of *L. pneumophila*. Amoebae exposed to *L. pneumophila* and containing adhered *L. pneumophila*, *L. pneumophila* antigens, or both, showed no increased pathogenic potential on intranasal inoculation of mice. Similarly, *L. pneumophila* propagated in chronically infected amoeba cultures showed no increase in virulence on intraperitoneal inoculation of guinea pigs relative to *L. pneumophila* grown in yeast extract broth. Photo-micrographs showed intracellular vacuoles containing *L. pneumophila* in the process of binary fission that was accompanied by alignment of mitochondria and ribosome-like structures along the vacuole membrane. Although these intracellular events are remarkably similar to that seen in *L. pneumophila* replication within human monocytes, an increase in bacteria was not observed (Newsome et al., 1985). Notably, the maximum numbers of culturable *L. pneumophila* cells grown within *N. fowleri* were 44 CFU per trophozoite (Buse and Ashbolt, 2012).

As *L. pneumophila* survive in aquatic environments in association with other bacteria, the effects of non-*Legionella* bacteria (*Escherichia coli*, *Aeromonas hydrophila*, *Flavobacterium breve*, and *Pseudomonas aeruginosa*) on the uptake of *L. pneumophila* by *Naegleria* have been studied. It was observed the non-*Legionella* bacteria did not compete with *L. pneumophila* for uptake, suggesting that the amoeba hosts took in *L. pneumophila* through a specific and presumably highly efficient uptake mechanism. These findings demonstrate the ability of *Naegleria* to serve as a proper host for the replication and distribution of *L.*

pneumophila (Declerck et al., 2005). Bacterial attachment and uptake occurred through a receptor-mediated endocytosis, which required *de novo* synthesis of host proteins. *Legionella pneumophila* showed a high affinity to GalNAcbeta1-4Gal domain of the *N*-acetyl-D-galactosamine receptor of *Naegleria* (Declerck et al., 2007).

Other knowledge of endosymbiont in *N. fowleri* is limited, however several lines of evidence suggest that endosymbionts play a key role in the biology of *N. gruberi* (Michel et al., 1999; 2000; Walochnik et al., 2005) (Figure 10.1 - 10.4) as well as support the growth of *Vibrio cholerae*. Using amoeba co-culture, Casson et al., (2008) isolated a *Naegleria* endosymbiont. Phenotypic, genetic, and phylogenetic analyses supported its affiliation as

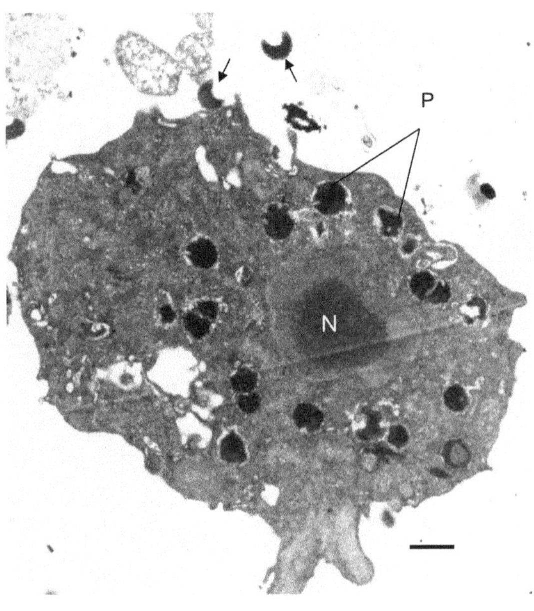

Figure 10.1. *N. lovaniensis* with *Protochlamydia naegleriophila*, strain KNic, as endoparasites (P), arrows free endoparastes, Bar is 1.0 μm (kindly provided by R. Michel and B. Hauröder, Central Institute of the Armed Forces Medical Service, Koblenz, Germany).

Protochlamydia naegleriophila sp. nov. that may have a role in pneumonia (Casson et al., 2008). (Figure 10.1). In microcosms pre-inoculated with trophozoites of amoebae, *V. cholerae* tested survived and multiplied during 24 h. In control microcosms without trophozoites of amoebae, survival of the *V. cholerae* was decreased. *V. cholerae* survive ingestion within amoebae and subsequent encystation (Thom et al., 1992). The high abundance in lakes, ponds and water distribution networks of these organisms indicate their potential symbiotic relationship, possibly contributing to survival and dissemination of these pathogens to humans and other animals.

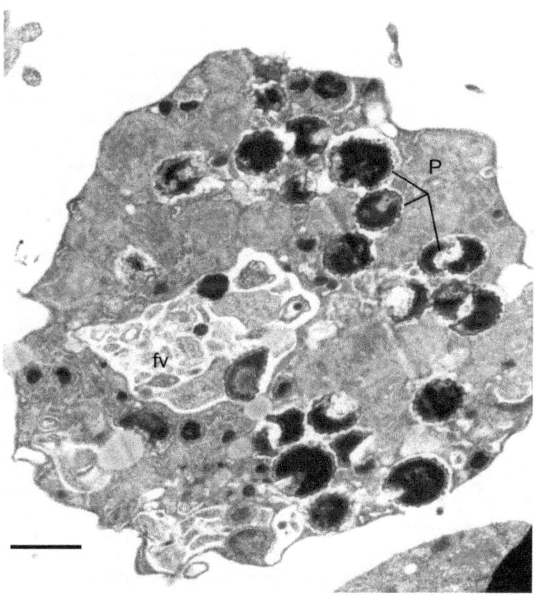

Figure 10.2. Numerous *P. naegleriophila* endoparasites (P) can be observed in the cytoplasm of the host *N. lovaniensis*. Bar is 1.0 µm (kindly provided by R. Michel and B. Hauröder, Central Institute of the Armed Forces Medical Service, Koblenz, Germany).

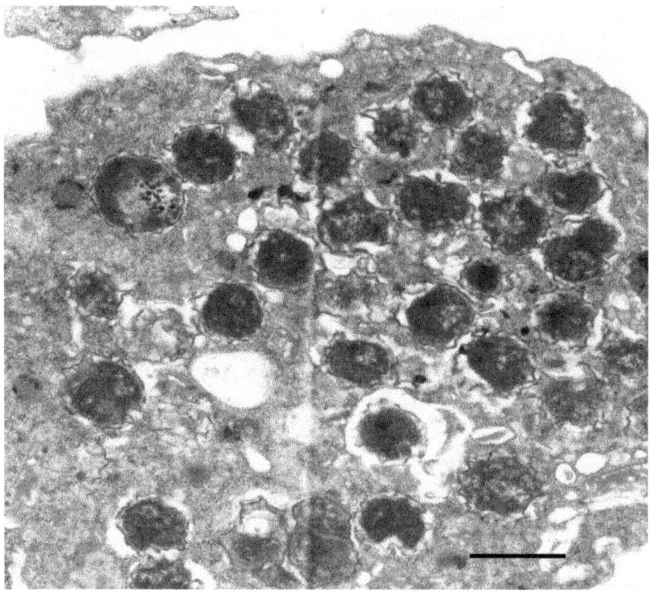

Figure 10.3. *N. clarki* harboring numerous *P. naegleriophila*-endoparasites "Pcb" within the cytoplasm. Bar is 1.0 μm (kindly provided by R. Michel and B. Hauröder, Central Institute of the Armed Forces Medical Service, Koblenz, Germany).

10.3. Bacterial evasion of predation by *Naegleria* spp.

Environmental bacteria are exposed to predators such as free-living amoeba including *Naegleria* spp. that influence their function and survival. The grazing activity of protozoan predators significantly impacts the dynamics, diversification, and evolution of bacterial communities in soil ecosystems. To evade protozoan predation, bacteria employ various defense strategies. Soil-dwelling *Pseudomonas fluorescens* produce the cyclic lipopeptide surfactants (CLPs) massetolide and viscosin, respectively, in a quorum-sensing-independent manner. CLP production protect these bacteria from protozoan (*Naegleria*) predation as, compared to CLP-deficient mutants. CLP-producing bacteria had a direct deleterious impact on the survival of *Naegleria*. *In vitro* assays further showed that *Naegleria* was three times more

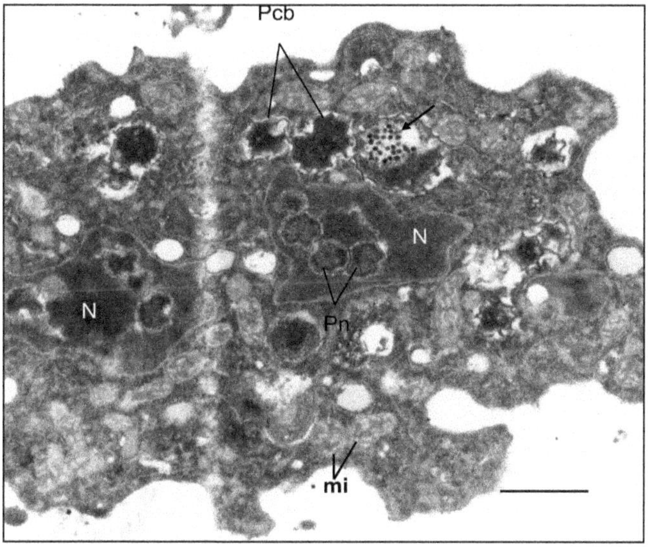

Figure 10.4. Trophozoite of *N. clarki* with two nuclei harboring two different populations of endocytobionts: 1. "Pcb" replicating within the cytolasm (*P. naegleriophila*), 2. "Pn" confined as endosymbiont to the karyosome and karyoplasm. Identification not yet published – in preparation with a group at Vienna/ Austria, N is nucleus, mi is mitochondria, arrow – unidentified bacteriophages differentiated within one chlamydia-like endoparasite, Pcb. Bar is 1.0 µm (kindly provided by R. Michel and B. Hauröder, Central Institute of the Armed Forces Medical Service, Koblenz, Germany).

sensitive to viscosin than to massetolide and that exposure of *P. fluorescens* to *Naegleria* resulted in upregulation of CLP biosynthesis genes. Enhanced expression of the massABC and viscABC genes did not require physical contact between the two organisms. These findings document a new natural function of CLPs and highlight that bacterium-protozoan interactions can result in activation of an antipredator response in prey populations (Mazzola et al., 2009).

Later studies showed that *Bacillus licheniformis* has remarkable amoebicidal activity against *Naegleria* sp. Physical-chemical characteristics, partial purification and biological activities of substance(s) produced by *B. licheniformis* showed that the activity is stable at high temperature (up to 100°C) and extremes of pH (2.5 - 9.5) and also at -20°C for months. Its production is greatly influenced by oxygenation of the cultures and is probably related to the sporulation process of the bacterium. Scanning electron microscope observations reveal that amoebae are lysed after a few minutes of contact with *B. licheniformis* (Cordovilla et al., 1993), however it was influenced by the composition of the medium and the initial amoeba/bacterium ratio. While a ratio of 50 caused complete lysis of amoebae in soil extract with 0.3% glucose before 72 h, this ratio had to be at least 12-fold lower in order to obtain similar results in Cline medium. Amoebicin production was stimulated by the presence of amoebae. Electron microscopy showed that amoebae co-cultivated in the Cline medium contained bacteria enclosed in digestive vacuoles, while amoebae from 0.3% glucose medium co-cultures did not (Lebbadi et al., 1995). Three amoebolytic peptides with amoebolytic activity have been purified from culture supernatants of *B. licheniformis* (amoebicins m4-A, m4-B, and m4-C). They are hydrophilic peptides consisting of six different amino acids (Asp, Glu, Ser, Thr, Pro, Tyr). Their molecular weights ranged from 3,000 to 3,200. Purified amoebicins were active against human pathogenic and non-pathogenic strains of *Naegleria*. They are heat- and protease-resistant. They also exhibit antibiotic action against yeasts (*Saccharomyces heterogenicus* and *Cryptococcus neoformans*) and several fungal species (*Aspergillus niger*, *Microsporum canis*, *Mucor plumbeus*, and *Trychophyton mentagrophytes*). Their antibacterial spectrum appears to be restricted to *Bacillus megaterium*, *Corynebacterium glutamicum*, and *Sarcina* sp. (Gálvez et al., 1993; Lebbadi et al., 1994). Amoebicin m4-A inhibited exerted bactericidal effects *Bacillus megaterium* by inhibiting protein, DNA, and RNA synthesis, and by depleting the membrane electrical potential of the bacterium. Synthesis of DNA and RNA by *N. fowleri* is also inhibited. Liposomes constructed from *L*-alpha-phosphatidylcholine become permeable to ions, low-molecular-weight solutes, and high-molecular-weight polymers after treatment with amoebicin (Lebbadi et al., 1994b).

Additionally, *B. licheniformis* produces three hydrophobic peptides (amoebicins d13-A, d13-B, and d13-C) that elicit antiamoebic activity against human-pathogenic and nonpathogenic species of *Naegleria* and have a broad spectrum of antibacterial activity. The three amoebicins have the same amino acid composition (three Asp, two Glu, two Val, and nine Leu residues) and molecular weight (1,870). Amoebicin d13-B causes lysis of amoebae through disorganization of the cell membrane. It also induces permeability to 86Rb and membrane disruption in asolectin vesicles (Gálvez et al., 1994).

Diffusates from dormant spores of *Aspergillus fumigatus* were found to be inhibitory to soil protozoa. The anti-amoebal effects of diffusates from clinical and environmental isolates of *A. fumigatus* and of *A. terreus* significantly inhibited the growth of *N. gruberi*, however their effects on *N. fowleri* remains undetermined (Hobson, 2000).

11

Conclusions and Future Studies

Abstract

The diagnosis of *N. fowleri* infections is not straightforward and treatment is problematic. Further research is needed for prevention, pathogenesis and therapy and in particular for the development of rapid, cost-effective, and non-invasive assays for early diagnosis. Moreover, epidemiology studies are needed to determine the true burden of *N. fowleri* infection, both in humans and animals, and whether disease patterns differ around the world. There is a need to determine the effect of environmental factors, demographic factors, socioeconomic factors, in the spread, transmission, and prevalence of the parasite, as well as host susceptibility in contracting *N. fowleri* infection. The precise molecular mechanisms underlying *N. fowleri* pathogenesis and the role of inflammation and immune response need to be elucidated, as well as the understanding of mechanisms of flagellation, encystation, excystation, and associated metabolic pathways. In addition, there is a need for improvement of educational efforts to raise awareness among the public as well as the medical community (neurologists, physicians) of this devastating infection.

11.1. Rapid and non-invasive diagnosis

Primary amoebic meningoencephalitis is difficult to differentiate from bacterial meningoencephalitis, leading to delays in correct diagnosis. Primary amoebic meningoencephalitis is suspected only if cerebrospinal fluid analysis indicates purulent meningitis without bacteria (da Rocha-Azevedo et al., 2009). The majority of patients have increased intracranial pressure, and under these circumstances, drawing the cerebrospinal fluid via lumbar puncture has its own complications such as herniation of the brain stem. In this case, the primary focus is to lower the intracranial pressure using drugs such as Mannitol, which in the presence of intracranial haemorrhage complicates the management of the patient (Visvesvara, 2010; Tilak et al., 2008). Under these circumstances, performing the lumbar puncture is difficult prior to lowering intracranial pressure. The natural route of amoebal entry into the brain is the nasal cavity, followed by migration along the olfactory neuroepithelium. Owing to the proximity of the cribriform plate to the inferior surface of the frontal lobe, this area is the first affected where trophozoites tend to proliferate prior to spreading throughout the central nervous system. It is proposed that secretions collected from the infracribrial part of the nasal cavity may serve as a valuable sample collection site to detect amoebae (Figure 11.1) (Baig and Khan, 2015). For example, motile trophozoites were discovered in nasal secretions of a primary amoebic meningoencephalitis patient as well as in the cerebrospinal fluid and identified as *N. fowleri* by culturing the parasites (Singh et al., 1998). These studies suggest that secretions collected from the nasal cavity offer a valuable sample collection site. The wet preparation of the lavaged fluid can be examined under the microscope or using PCR-based assays to detect *N. fowleri*. It was hypothesized that the possibility of obtaining the parasite from such a flushing is high because the olfactory mucosa and the bulb that overlies the cribriform bone is a natural route of the infection by *N. fowleri*. In addition, the proposed strategy would bypass the complications of cerebrospinal fluid collection via invasive lumbar puncture and delays associated with lowering intracranial pressure. The usefulness of the proposed strategy in the early diagnosis of *N. fowleri* requires further research.

11.2. Antiamoebic anesthetic agents

The anesthetic agents, haloperidol and loperamide show anti-*N. fowleri* effects (Baig et al., 2014). Given that primary amoebic meningoencephalitis patients suffer repeated seizures and pain, it has been proposed that the aforementioned agents should be added to complement

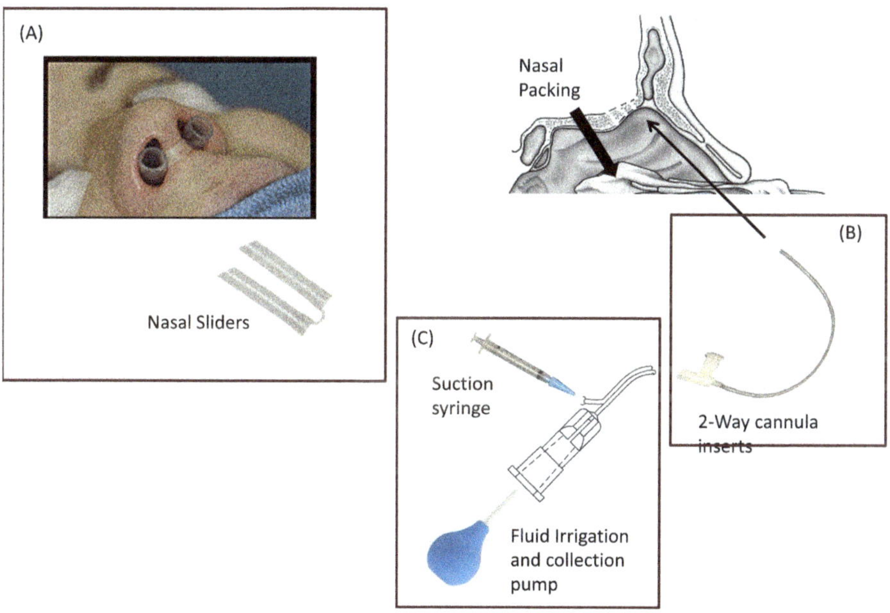

Figure 11.1. Device for mucosal secretion collection for *N. fowleri* diagnosis. The proposed device is made of three components, **(A)** nasal sliders, **(B)** two way insert cannula, and **(C)** an irrigation and collection component. Saline will be used to irrigate the upper nasal cavity in the region of the olfactory mucosa, and subsequently, secretions will be collected passively back into the pump; the remaining secretions can be manually drawn using a suction syringe. The collected sample would be tested for the presence of *N. fowleri*. To avoid aspiration into the lungs, posterior nasal packing could be used ahead of nasal irrigation (from Acta Tropica, 2015, 142:86-88).

therapy by exhibiting amoebicidal effects. If neuroleptanalgesia is administered by inhalation, some topical effects are expected around the cribriform plate, olfactory bulb, and frontal lobe of the brain because they would be en-route of administration to the central nervous system, thus effectively targeting the epicenter of trophozoite proliferation. In this regard, a 'transcribrial route' device could provide an effective route of delivery for the vaporized drugs, including anesthetic agents as well as known anti-*N. fowleri* compounds (Baig and Khan 2015). The proposed delivery system would outweigh the current intraventricular, intrathecal routes that require surgical procedures and have their own attendant complications. Furthermore, the current drug delivery routes result in dilution of the drugs in the plasma or cerebrospinal fluid before reaching the inferior surface of the frontal lobe of the brain. Even if drugs are injected into cerebrospinal fluid, drainage into the venous blood by arachnoid villi would reduce the concentration of the drug. Further research is needed to determine the effectiveness of the aforementioned diagnostic, anesthetic, and treatment strategies to help improve patient compliance as well as the morbidity and mortality of this devastating infection.

11.3. Drug delivery

The transcribrial route (intranasal route) has been described as a portal for administering drugs against primary amoebic meningoencephalitis due to *N. fowleri* (Baig and Khan, 2014). The proposed intranasal route can deliver drugs across the cribriform plate (an anatomically porous bone) that is located at the roof of the nasal cavity (Figure 11.2). The underlying rationale for proposing such a portal is that many therapeutically-desired drugs like amphotericin B fail to attain minimum inhibitory concentration in the central nervous system, when administered systemically. The proposed route will be advantageous in that it will (i) bypass selectivity of the blood-brain barrier that limits drug permeability to the brain tissue, (ii) trail the natural route of entry of *N. fowleri* into the brain, (iii) target the site of infection, (iv) exert lethal effect of drugs by attaining minimum inhibitory concentration at a lower dose without venous drainage, (v) avoid adverse effects due to systemic administration. Olfactory targeting through intranasal delivery will provide access to the central nervous system without affecting the integrity of the blood-brain barrier. It is argued that the proposed route will enable the use of water soluble drugs to attain minimum inhibitory concentration at the epicenter of infection, overcome the blood-brain barrier hindrance, and exhibit minimal adverse effects with the benefits of topical administration and dose adjustment at a faster pace. Given that primary amoebic meningoencephalitis patients are receiving adjuvant therapies including anti-seizure, strong hypnotics, and even general anesthetics, patient compliance with the proposed route should not pose any concern. Further research is needed to test the effectiveness of the proposed route to

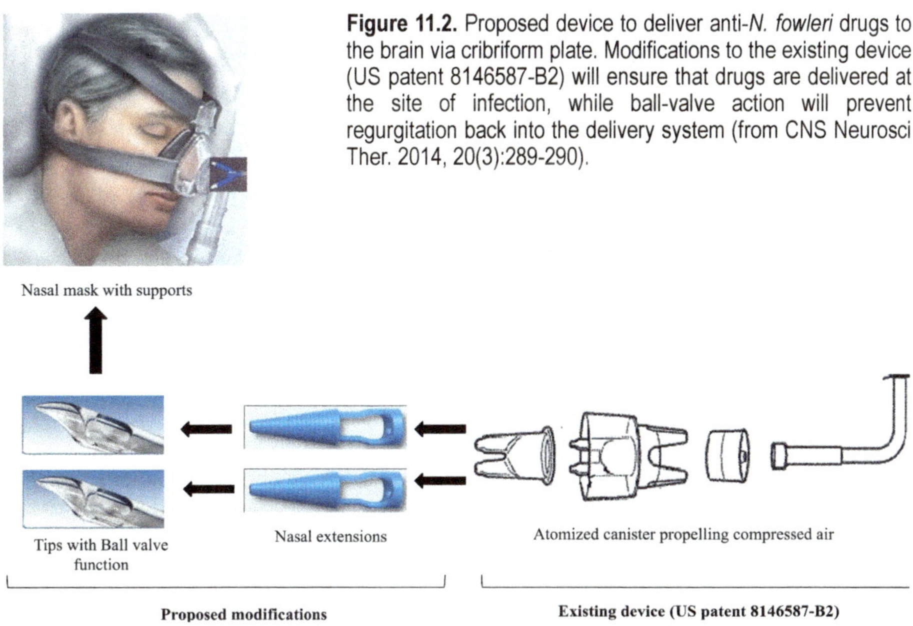

Figure 11.2. Proposed device to deliver anti-*N. fowleri* drugs to the brain via cribriform plate. Modifications to the existing device (US patent 8146587-B2) will ensure that drugs are delivered at the site of infection, while ball-valve action will prevent regurgitation back into the delivery system (from CNS Neurosci Ther. 2014, 20(3):289-290).

administer drugs in the management of primary amoebic meningoencephalitis due to *N. fowleri*.

11.4. Drug repurposing

Testing of drugs that are clinically-approved against various diseases with known mechanisms of action and targets/ biochemical pathways offers research avenues as the pharmacokinetics profiles, safety margins etc. have been already established. Among the various FDA-approved drugs tested against *N. fowleri*, digoxin and procyclidine were shown to possess potent amoebicidal effects (Baig et al., 2014). In view of the devastating nature of this disease and the problems associated with chemotherapy, this approach can offer promising drug-leads for clinical practice. Once a number of potential compounds are identified *in vitro*, with known pharmacokinetics, and pharmacodynamics profiles, the findings can lead to *in vivo* testing at a faster pace. The drug discovery of effective anti-*N. fowleri* drugs over the last few decades has been disappointing, largely due to neglect by the funding agencies. The fact that PAM is a rare disease in developed countries and affects communities mostly in developing countries, where millions of people are dying due to other diseases, makes it logical to undertake drug repurposing approach in our search for effective anti-*N. fowleri* drugs.

Potential opportunities for drug repurposing by developing an automated approach to pre-screen the predicted proteomes of parasites against databases of known drug targets has been proposed. A combination of Ruby scripts that leverage data from the DrugBank and ChEMBL databases, MySQL, and BLAST to predict potential drugs and their targets from parasite published genomes. Although proteome-wide screens for drug targets have limilations, in-silico methods can be developed that are fast, broad, inexpensive, and effective. These methods can be used to construct a library for more directed small molecule screening, or pipelined into structural modeling and docking programs to facilitate target-based drug development (Sateriale et al., 2014).

11.5. Biomarkers

Absence of distinct biomarker(s) of PAM in the mucosal secretions, blood, CSF makes it difficult to study the impact of interventions against it. Biomarkers for PAM have been challenging to find, partly because of our inadequate understanding of its pathogenesis and pathophysiology. Thus there is a need to identify novel biomarkers for PAM. Using analytical tools such as mass-spectrometry, NMR etc., there is also a need to study and compare biochemical profiles of pathogenenic and non-pathogenic *Naegleria* as well as patients from different populations, gender, ethinicity to identify potential predisposing factors. The paucity of PAM studies during the last few decades is partly to blame for the lack of understanding of the pathogenesis and pathophysiology of PAM, and the subsequent lack of its good biomarkers. Major technological advances in studying pathophysiologic mechanisms in general have not been applied to the study of PAM so far. If we are to identify sensitive and specific biomarkers of PAM, we must first understand the pathophysiologic mechanisms that underlie this disease. The reasons for the lack of aforementioned studies are multifactorial. An important issue is that PAM is rare and/or occurs in resource-limited settings, which may not have the infrastructure to perform such experimentation or have other research priorities of health-related problems of major significance. There, clearly, is a need to identify and test new promising candidate biomarkers of PAM in mucosal secretions, blood, and CSF, and then correlating them with the histopathologic features of affected patients, which will provide a mechanistic basis of PAM.

11.6. Drug targets

A complete understanding of the biology and pathogenesis of *N. fowleri* will identify molecules that could be of value to develop therapeutic interventions. Ondarza (2007) proposed that the exploration of several pathways can reveal drug targets, including, a) hydrolytic enzymes which are secreted by amoebae to invade the human host, b) glycolytic enzymes that differ from the host enzyme, c) thiols and enzymes of redox metabolism, that maintains the reducing environment within the cell, d) antioxidant enzymes to regulate the oxidative stress, e) enzymes for the synthesis of trypanothione, f) some of the proteins that assemble the secretory vesicles with the cell membrane, and g) encystation pathways and cyst-wall assembly proteins.

11.7. A model organism with pathogenic potential

The biology of *N. fowleri* is fascinating (Figure 11.3). With its ability to switch phenotypes to an active trophozoite form to grow vegetatively, or transform into a flagellate to travel long distance, or switch into a dormant cyst form to endure harsh conditions, as well as the fact that it is a single-celled eukaryotic protozoan, it is an excellent organism for the study of biochemistry, of cellular microbiology, molecular mechanisms of cell differentiation, physiology, cellular interactions with other microbes or human cells, phagocytosis, flagellar and amoebal motility, and evolutionary studies. Our understanding of *N. fowleri* pathogenesis, both at the molecular-level and at the cellular-level, as well as its ability to transmit, adapt to diverse conditions, overcome host

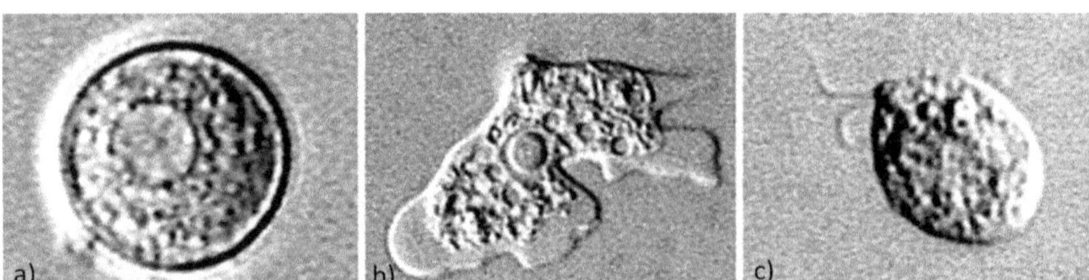

Figure 11.3. *N. fowleri* (a) cyst form, and (b) trophozoite form, and (c) flagellate form, viewed under phase contrast microscopy. X1000. (courtesy: G. S. Visvesvara, CDC, USA).

barriers and emerge as an infective trophozoite will provide targets for therapeutic interventions.

In conclusion, at present, the diagnosis of the disease is not straightforward and treatment is problematic. In view of the devastating nature of the disease and the problems associated with the therapy, our future goals are to find means to identify individuals who are at risk and to provide them with rationally-designed strategies to protect against contracting the infection. The preventative strategies may vary among different communities, from the use of chlorine or filters to decontaminate water, to the use of nose clips or use of clean water for nasal cleansing / irrigation. It is known that adhesion of the trophozoite to nasal mucosa is the first step in parasite infection, and induction of mucosal immunity is another avenue to counter infection progression. Several virulence factors have been described, some of which are potential vaccine targets for the susceptible population. The availability of *in vitro* and *in vivo* models, the genome information and advances in genomics, proteomics and bioinformatics will expedite the identification of novel drug targets. Given the rarity of the disease, chemotherapy will remain the primary choice to develop therapeutic interventions. Many of the drugs tested target the functional aspects of *N. fowleri*, as it is "easier to raze function of an organisms than its structure". However, there are downsides to this approach. Being a eukaryote, *N. fowleri* share functional homologies to mammalian cells. Consequently, many of the available drugs cannot be prescribed at effective concentrations due to their unwanted side effects. This is particularly relevant for *N. fowleri* infection, as drugs are given intravenously, and expected to cross the blood-brain barrier to access the central nervous system, to target the intracerebral parasite. In this process, drugs target all tissues, and can affect their physiology before reaching the target site, at effective concentration. Hence, there is a need to develop a targeted therapeutic approach, or identify drugs that can affect *N. fowleri* viability without affecting the host cells. Future research is needed to identify drugs and/or chemotherapeutic approaches of potential value. A better understanding of drugs, their targets, and mechanisms of action will facilitate the development of more effective chemotherapies against this rare, but fatal infection.

There is an urgent need for continued research in the development of strategies in the prevention, pathogenesis and therapy for *N. fowleri* infections. There is a need to develop rapid, cost-effective, and non-invasive assays in the early diagnosis of *N. fowleri* infection, and to demonstrate their efficacy in clinical practice. Moreover, epidemiology studies are needed to determine the true burden of *N. fowleri* infection, both in humans and animals, and whether disease patterns differ around the world. There is a need to determine the effect of environmental factors, demographic factors, socioeconomic factors, in the spread, transmission, and prevalence of the parasite, as well as host susceptibility in contracting *N. fowleri* infection. The precise molecular mechanisms underlying *N. fowleri* pathogenesis and the role of inflammation and immune response need to be elucidated, as well as the understanding of mechanisms of flagellation, encystation, excystation, and associated metabolic pathways will help identify potential therapeutic targets, and/or help develop targeted anti-*N. fowleri* chemotherapeutic and immunotherapeutic approaches. In addition, there is a need for improvement of educational efforts to raise awareness among the public as well as the medical community (neurologists, physicians) of this devastating infection.

References

Abraham, S.N., and Lawande, R.V. (1982). Incidence of free-living amoebae in the nasal passages of local population in Zaria, Nigeria. J. Trop. Med. Hyg. 85, 217-22.

Adams, A.C., John, D.T., and Bradley, S.G. (1976). Modification of resistance of mice to *Naegleria fowleri* infections. Infect. Immun. 13, 1387-1391.

Adams, M., Andrews, R.H., Robinson, B., Christy, P., Baverstock, P.R., Dobson, P.J., and Blackler, S.J. (1989). A genetic approach to species criteria in the amoeba genus *Naegleria* using allozyme electrophoresis. Int. J. Parasitol. 19, 823-834.

Adl, S.M., Simpson, A.G., Farmer, M.A., Andersen, R.A., Anderson, O.R., Barta, J.R., Bowser, S.S., Brugerolle, G., Fensome, R.A., Fredericq, S., James, T.Y., Karpov, S., Kugrens, P., Krug, J., Lane, C.E., Lewis, L.A., Lodge, J., Lynn, D.H., Mann, D.G., McCourt, R.M., Mendoza, L., Moestrup, O., Mozley-Standridge, S.E., Nerad, T.A., Shearer, C.A., Smirnov, A.V., Spiegel, F.W., and Taylor, M.F. (2005). The new higher level classification of eukaryotes with emphasis on the taxonomy of protists. J. Eukaryot. Microbiol. 52, 399-451.

Ahmad, A.F., Lonnen, J., Andrew, P.W., and Kilvington, S. (2011). Development of a rapid DNA extraction method and one-step nested PCR for the detection of *Naegleria fowleri* from the environment. Water Res. 45, 5211-5217.

Ahn, M.H., and Im, K.I. (1984). Experimental meningo-encephalitis by *Naegleria fowleri* in mice. Kisaengchunghak Chapchi. 22, 253-258.

Ahn, M.H., and Min, D.Y. (1989). Resistance to *Naegleria fowleri* infection passively acquired from immunized splenocyte, serum or milk. Kisaengchunghak Chapchi. 27, 79-86.

Aldape, K., Huizinga, H., Bouvier, J., and McKerrow, J. (1994). *Naegleria fowleri*: characterization of a secreted histolytic cysteine protease. Exp. Parasitol. 78, 230-241.

Alonso, P., and Zubiaur, E. (1985). Phagocytic activity of three *Naegleria* strains in the presence of erythrocytes of various types. J. Protozool. 32, 661-664.

Anderson, K., and Jamieson, A. (1972). Primary amoebic meningoencephalitis. Lancet 2(7773):379.

Anderson, K., and Jamieson, A. (1972). Agglutination test for the investigation of the genus *Naegleria*. Pathol. 4, 273-278.

Anderson, K., and Jamieson, A. (1974). Bacterial suspensions for the growth of *Naegleria* species. Pathol. 6, 79-84.

Annonymous (2013). Notes from the field: primary amoebic meningoencephalitis associated with ritual nasal rinsing--St. Thomas, U.S. Virgin islands, 2012. Morb. Mortal. Wkly. Rep. 62, p. 903.

Antonios, S.N. (2010). Scanning electron microscopic study of trophozoite and cyst stages of *Naegleria fowleri*. J. Egypt. Soc. Parasitol. 40, 271-276.

Apley, J., Clarke, S.K., Roome, A.P., Sandry, S.A., Saygi, G., Silk, B., and Warhurst, D.C. (1970). Primary amoebic meningoencephalitis in Britain. Br. Med. J. 1(5696): 596-599.

Araujo, F.G., Shepard, R.M., and Remington, J.S. (1991). *In vivo* activity of the macrolide antibiotics azithromycin, roxithromycin and spiramycin against *Toxoplasma gondii*. Eur. J. Clin. Microbiol. Infect. Dis. 10, 519-524.

Averner, M., and Fulton, C. (1966). Carbon dioxide: signal for excystment of *Naegleria gruberi*. J. Gen. Microbiol. 42, 245-255.

Baek, I.K., Chung, S., Suh, M.R., Hwang, D.S., Kang, D., and Lee, J. (2012). Coordinate synthesis but discrete localization of homologous N-glycosylated proteins, CLP and CLB, in *Naegleria pringsheimi* flagellates. J. Eukaryot. Microbiol. 59, 614-624.

Baig, A.M., and Khan, N.A. (2014). Novel chemotherapeutic strategies in the management of primary amoebic meningoencephalitis due to *Naegleria fowleri*. CNS Neurosci. Therap. 20, 289-290.

Baig, A.M., Kulsoom, H., Khan, N.A. (2014). Primary amoebic meningoencephalitis: amoebicidal effects of clinically approved drugs against *Naegleria fowleri*. J. Med. Microbiol. 63, 760-762.

Baig, A.M., and Khan, N.A. (2015). Tackling infection owing to brain-eating amoeba. Acta Trop. 142, 86-88.

Band, R.N., and Balamuth, W. (1974). Hemin replaces serum as a growth requirement for *Naegleria*. Appl. Microbiol. 28, 64-65.

Barbeau, J., and Buhler, T. (2001). Biofilms augment the number of free-living amoebae in dental unit waterlines. Res. Microbiol. 152, 753-760.

Barbour, S.E., and Marciano-Cabral, F. (2001). *Naegleria fowleri* amoebae express a membrane-associated calcium-independent phospholipase A(2). Biochim. Biophys. Acta. 1530, 123-133.

Baverstock, P.R., Illana, S., Christy, P.E., Robinson, B.S., and Johnson, A.M. (1989). srRNA evolution and phylogenetic relationships of the genus *Naegleria* (Protista: Rhizopoda). Mol. Biol. Evol. 6, 243-257.

Behets, J., Seghi, F., Declerck, P., Verelst, L., Duvivier, L., Van Damme, A., and Ollevier, F. (2003). Detection of *Naegleria* spp. and *Naegleria fowleri*: a comparison of flagellation tests, ELISA and PCR. Water Sci. Technol. 47, 117-122.

Behets, J., Declerck, P., Delaedt, Y., Verelst, L., and Ollevier, F. (2006). Quantitative detection and differentiation of free-living amoeba species using SYBR green-based real-time PCR melting curve analysis. Curr. Microbiol. 53, 506-509.

Behets, J., Declerck, P., Delaedt, Y., Verelst, L., and Ollevier, F. (2007). A duplex real-time PCR assay for the quantitative detection of *Naegleria fowleri* in water samples. Water Res. 41, 118-126.

Belofsky, G., Carreno, R., Goswick, S.M., and John, D.T. (2006). Activity of isoflavans of *Dalea aurea* (Fabaceae) against the opportunistic amoeba *Naegleria fowleri*. Planta Med. 72, 383-386.

Biddick, C.J., Rogers, L.H., and Brown, T.J. (1984). Viability of pathogenic and nonpathogenic free-living amoebae in long-term storage at a range of temperatures. Appl. Environ. Microbiol. 48, 859-860.

Blackler, S.J., and Sommerville, R.I. (1988). Carbonic acid and the excystment of *Naegleria gruberi* (Heterolobosea: Vahlkampfiidae). Int. J. Parasitol. 18, 785-791.

Bok, J., Jin, Y., and Lee, J. (1995). Independent mechanisms are utilized for the coordinate and transient accumulation of two differentiation-specific mRNAs during differentiation of *Naegleria gruberi* amoebae into flagellates. Exp. Cell Res. 219, 47-53.

Bols, N.C., Corff, S., and Yuyama, S. (1977). Rapid decline thymidine kinase activity during differentiation of *Naegleria*. J. Cell. Physiol. 90,271-280.

Bose, K., Ghosh, D.K., and Bhattacharya, A. (1989). Membrane carbohydrate characterization of *Acanthamoeba astronyxis*, *A. castellanii* and *Naegleria fowleri* by fluorescein-conjugated lectins. Int. J. Parasitol. 19, 737-741.

Bose, K., Ghosh, D.K., Ghosh, K.N., Bhattacharya, A., and Das, S.R. (1990). Characterization of potentially pathogenic free-living amoebae in sewage samples of Calcutta, India. Braz. J. Med. Biol. Res. 23, 1271-1278.

Bradley, S.G., Toney, D.M., Zhang, Y., and Marciano-Cabral, F. (1996). Dependence of growth, metabolic expression, and pathogenicity of *Naegleria fowleri* on exogenous porphyrins. J. Parasitol. 82, 763-768.

Brinkley, C., and Marciano-Cabral, F. (1992). A method for assessing the migratory response of *Naegleria fowleri* utilizing [3H]uridine-labeled amoebae. J. Protozool. 39, 297-303.

Brown, T. (1978). Observations by light microscopy on the cytopathogenicity of *Naegleria fowleri* in mouse embryo-cell cultures. J. Med. Microbiol. 11, 249-259.

Brown, T. (1979). Observations by immunofluorescence microscopy and electron microscopy on the cytopathogenicity of *Naegleria fowleri* in mouse embryo-cell cultures. J. Med. Microbiol. 12, 363-371.

Brown, T. (1979b). Inhibition by amoeba-specific antiserum and by cytochalasin B of the cytopathogenicity of *Naegleria fowleri* in mouse embryo-cell cultures. J. Med. Microbiol. 12, 355-362.

Brown, T.J., Cursons, R.T., and Keys, E.A. (1982). Amoebae from antarctic soil and water. Appl. Environ. Microbiol. 44, 491-493.

Budge, P.J., Lazensky, B., Van Zile, K.W., Elliott, K.E., Dooyema, C.A., Visvesvara, G.S., Beach, M.J., and Yoder, J.S. (2013). Primary amoebic meningoencephalitis in Florida: a case report and epidemiological review of Florida cases. J. Environ. Health 75, 26-31.

Burnette-Curley, D., Marciano-Cabral, F., Fischer-Stenger, K., and Cabral, G.A. (1993). delta-9-Tetrahydrocannabinol inhibits cell contact-dependent cytotoxicity of Bacillus Calmétte-Guérin-activated macrophages. Int. J. Immunopharmacol. 15, 371-382.

Burri, D.C., Gottstein, B., Zumkehr, B., Hemphill, A., Schürch, N., Wittwer, M., and Müller, N. (2012). Development of a high- versus low-pathogenicity model of the free-living amoeba *Naegleria fowleri*. Microbiol. 158, 2652-2660.

Buse, H.Y., and Ashbolt, N.J. (2012). Counting *Legionella* cells within single amoeba host cells. Appl. Environ. Microbiol. 78, 2070-2072.

Bush, L.E., and John, D.T. (1988). Intranasal immunization of mice against *Naegleria fowleri*. J. Protozool. 35, 172-176.

Butt, C.G., Baro, C., and Knorr, R.W. (1968). *Naegleria* (sp.) identified in amoebic encephalitis. Am. J. Clin. Pathol. 50, 568-574.

Byers, T.J. (1986). Molecular biology of DNA in *Acanthamoeba*, *Amoeba*, *Entamoeba*, and *Naegleria*. Int. Rev. Cytol. 99, 311-341.

Cable, B.L., and John, D.T. (1986). Conditions for maximum enflagellation in *Naegleria fowleri*. J. Protozool. 33, 467-472.

Capewell, L.G., Harris, A.M., Yoder, J.S., Cope, J.R., Eddy, B.A., Roy, S.L., Visvesvara, G.S., Fox, L.M., and Beach, M.J. (2014). Diagnosis, clinical course, and treatment of primary amoebic meningoencephalitis in the United States, 1937-2013. J. Pediat. Infect. Dis. Soc. 4(4):e68-75.

Cariou, M.L., and Pernin, P. (1987). First evidence for diploidy and genetic recombination in free-living amoebae of the genus *Naegleria* on the basis of electrophoretic variation. Genetics. 115, 265-270.

Carrasco-Yepez, M., Rojas-Hernandez, S., Rodriguez-Monroy, M.A., Terrazas, L.I., and Moreno-Fierros, L. (2010). Protection against *Naegleria fowleri* infection in mice immunized with Cry1Ac plus amoebic lysates is dependent on the STAT6 Th2 response. Parasite Immunol. 32, 664-670.

Carrasco-Yepez, M., Campos-Rodriguez, R., Godinez-Victoria, M., Rodriguez-Monroy, M.A., Jarillo-Luna, A., Bonilla-Lemus, P., De Oca, A.C., and Rojas-Hernandez, S. (2013). *Naegleria fowleri* glycoconjugates with residues of α-D-mannose are involved in adherence of trophozoites to mouse nasal mucosa. Parasitol. Res. 112, 3615-3625.

Carrasco-Yepez, M., Campos-Rodriguez, R., Lopez-Reyes, I., Bonilla-Lemus, P., Rodriguez-Cortes, A.Y., Contis-Montes de Oca, A., Jarillo-Luna, A., Miliar-Garcia, A., and Rojas-Hernandez, S. (2014). Intranasal coadministration of Cholera toxin with amoeba lysates modulates the secretion of IgA and IgG antibodies, production of cytokines and expression of pIgR in the nasal cavity of mice in the model of *Naegleria fowleri* meningoencephalitis. Exp. Parasitol. 145 Suppl:S84-92.

Carter, R.F. (1969). Sensitivity to amphotericin B of a *Naegleria* sp. isolated from a case of primary amoebic meningoencephalitis. J. Clin. Pathol. 22, 470-474.

Carter, R.F. (1970). Description of a *Naegleria* sp. isolated from two cases of primary amoebic meningoencephalitis, and of the experimental pathological changes induced by it. J. Pathol. 100, 217-244.

Casson, N., Michel, R., Müller, K.D., Aubert, J.D., and Greub, G. (2008). *Protochlamydia naegleriophila* as etiologic agent of pneumonia. Emerg. Infect. Dis. 14, 168-172.

Ccaronerva, L. (1969). Amoebic meningoencephalitis: axenic culture of *Naegleria*. Science. 163(3867):576.

Cerva, L., Zimák, V., and Novák, K. (1969). Amoebic meningoencephalitis: a new amoeba isolate. Science. 163(3867):575-576.

Cerva, L. (1978). Some further characteristics of the growth of *Naegleria fowleri* and *N. gruberi* in axenic culture. Folia Parasitol. (Praha). 25, 1-8.

Cerva, L. (1980). *Naegleria fowleri*: trimethoprim sensitivity. Science 209(4464):1541.

Cerva, L., Jecná, P., and Hyhlík, R. (1980). *Naegleria fowleri* from a canal draining cooling water from a factory. Folia Parasitol. (Praha). 27, 103-107.

Cerva, L., Kasprzak, W., and Mazur, T. (1982). *Naegleria fowleri* in cooling waters of power plants. J. Hyg. Epidemiol. Microbiol. Immunol. 26, 152-161.

Cerva, L. (1986). *Naegleria fowleri* and *N. lovaniensis*: differences in sensitivity to trimethoprim and other antifolates. Z. Parasitenkd. 72, 585-590.

Cerva, L. (1989). *Acanthamoeba culbertsoni* and *Naegleria fowleri*: occurrence of antibodies in man. J. Hyg. Epidemiol. Microbiol. Immunol. 33, 99-103.

Cervantes-Sandoval, I., Serrano-Luna Jde, J., García-Latorre, E., Tsutsumi, V., and Shibayama, M. (2008). Characterization of brain inflammation during primary amoebic meningoencephalitis. Parasitol. Int. 57, 307-313.

Cervantes-Sandoval, I., Serrano-Luna Jde, J., García-Latorre, E., Tsutsumi, V., and Shibayama, M. (2008). Mucins in the host defence against *Naegleria fowleri* and mucinolytic activity as a possible means of evasion. Microbiol. 154, 3895-3904.

Cervantes-Sandoval, I., Serrano-Luna Jde, J., Meza-Cervantez, P., Arroyo, R., Tsutsumi, V., and Shibayama, M. (2009). *Naegleria fowleri* induces MUC5AC and pro-inflammatory cytokines in human epithelial cells via ROS production and EGFR activation. Microbiol. 155, 3739-3747.

Cervantes-Sandoval, I., Jesús Serrano-Luna, J., Pacheco-Yépez, J., Silva-Olivares, A., Tsutsumi, V., and Shibayama, M. (2010). Differences between *Naegleria fowleri* and *Naegleria gruberi* in expression of mannose and fucose glycoconjugates. Parasitol. Res. 106, 695-701.

Cetin, N., and Blackall, D. (2012). *Naegleria fowleri meningoencephalitis*. Blood 119(16), 3658.

Chang, S.L. (1958). Cultural, cytological and ecological observations on the amoeba stage of *Naegleria gruberi*. J. Gen. Microbiol. 18, 565-578.

Chang, S.L. (1958). Cytological and ecological observations on the flagellate transformation of *Naegleria gruberi*. J. Gen. Microbiol. 18, 579-585.

Chang, S.L. (1978). Resistance of pathogenic *Naegleria* to some common physical and chemical agents. Appl. Environ. Microbiol. 35, 368-375.

Chang, S.L. (1979). Pathogenesis of pathogenic *Naegleria* amoeba. Folia Parasitol. (Praha). 26, 195-200.

Chávez-Munguía, B., Omaña-Molina, M., Castañón, G., Bonilla, P., González-Lázaro, M., Hernández-Martínez, D., Salazar-Villatoro, L., Esparza-García, A., Martínez-Palomo, A., and Ortega-Pierres, G. (2009). Ultrastructural study of the encystation and excystation processes in *Naegleria* sp. J. Eukaryot. Microbiol. 56, 66-72.

Chávez-Munguía, B., Segovia-Gamboa, N., Salazar-Villatoro, L., Omaña-Molina, M., Espinosa-Cantellano, M., and Martínez-Palomo, A. (2011). *Naegleria fowleri*: enolase is expressed during cyst differentiation. J. Eukaryot. Microbiol. 58, 463-468.

Chávez-Munguía, B., Villatoro, L.S., Omaña-Molina, M., Rodríguez-Monroy, M.A., Segovia-Gamboa, N., and Martínez-Palomo, A. (2014). *Naegleria fowleri*: contact-dependent secretion of electrondense granules (EDG). Exp. Parasitol. 142, 1-6.

Cho, M.S., Jung, S.Y., Park, S., Kim, K.H., Kim, H.I., Sohn, S., Kim, H.J., Im, K.I., and Shin, H.J. (2003). Immunological characterizations of a cloned 13.1-kilodalton protein from pathogenic *Naegleria fowleri*. Clin. Diagn. Lab. Immunol. 10, 954-959.

Chu, D.M., Ferguson, T.J., and Marciano-Cabral, F. (2000). Protein kinase activation and protein phosphorylation in *Naegleria fowleri* amoebae in response to normal human serum. J. Eukaryot. Microbiol. 47, 40-47.

Chu, D.M., Woodward, J., Fritzinger, A., and Marciano-Cabral, F. (2002). Calcium-dependent protection from complement lysis in *Naegleria fowleri* amoebae. Cell. Calcium. 31, 105-114.

Chung, S., Cho, J., Cheon, H., Paik, S., and Lee, J. (2002). Cloning and characterization of a divergent alpha-tubulin that is expressed specifically in dividing amoebae of *Naegleria gruberi*. Gene 293, 77-86.

Chung, S., Kang, S., Paik, S., and Lee, J. (2007). NgUNC-119, *Naegleria* homologue of UNC-119, localizes to the flagellar rootlet. Gene 389, 45-51.

Clark, C.G., and Cross, G.A. (1987). rRNA genes of *Naegleria gruberi* are carried exclusively on a 14-kilobase-pair plasmid. Mol. Cell. Biol. 7, 3027-3031.

Clark, C.G., and Cross, G.A. (1988). Small-subunit ribosomal RNA sequence from *Naegleria gruberi* supports the polyphyletic origin of amoebas. Mol. Biol. Evol. 5, 512-518.

Clark, C.G., Cross, G.A., and De Jonckheere, J.F. (1989). Evaluation of evolutionary divergence in the genus *Naegleria* by analysis of ribosomal DNA plasmid restriction patterns. Mol. Biochem. Parasitol. 34, 281-296.

Clark, C.G. (1990). Genome structure and evolution of *Naegleria* and its relatives. J. Protozool. 37, 2S-6S.

Clark, C.G., Lai, E.Y., Fulton, C., and Cross, G.A. (1990). Electrophoretic karyotype and linkage groups of the amoeboflagellate *Naegleria gruberi*. J. Protozool. 37, 400-408.

Cleary, S.F., and Marciano-Cabral, F. (1986). Activated macrophages demonstrate direct cytotoxicity, antibody-dependent cellular cytotoxicity, and enhanced binding of *Naegleria fowleri* amoebae. Cell. Immunol. 98, 125-136.

Cleary, S.F., and Marciano-Cabral, F. (1986). Soluble amoebicidal factors mediate cytolysis of *Naegleria fowleri* by activated macrophages. Cell. Immunol. 101, 62-71.

Cline, M., Marciano-Cabral, F., and Bradley, S.G. (1983). Comparison of *Naegleria fowleri* and *Naegleria gruberi* cultivated in the same nutrient medium. J. Protozool. 30, 387-391.

Cline, M., Carchman, R., and Marciano-Cabral, F. (1986). Movement of *Naegleria fowleri* stimulated by mammalian cells *in vitro*. J. Protozool. 33, 10-13.

Cogo, P.E., Scagli, M., Gatti, S., Rossetti, F., Alaggio, R., Laverda, A.M., Zhou, L., Xiao, L., and Visvesvara, G.S. (2004). Fatal *Naegleria fowleri* meningoencephalitis, Italy. Emerg. Infect. Dis. 10, 1835-1837.

Cooke, D.W., Lallinger, G.J., Durack, D.T. (1987). *In vitro* sensitivity of *Naegleria fowleri* to qinghaosu and dihydroqinghaosu. J. Parasitol. 73, 411-413.

Cope, J.R., Conrad, D.A., Cohen, N., Cotilla, M., DaSilva, A., Jackson, J., Visvesvara, G.S. (2016). Use of the novel therapeutic agent miltefosine for the treatment of primary amebic meningoencephalitis: Report of 1 fatal and 1 surviving case. Clin. Infect. Dis. 62, 774-776.

Cordovilla, P., Valdivia, E., Gonzalez-Segura, A., Galvez, A., Martinez-Bueno, M., and Maqueda, M. (1993). Antagonistic action of the bacterium Bacillus licheniformis M-4 toward the amoeba *Naegleria fowleri*. J. Eukaryot. Microbiol. 40, 323-328.

Corff, S., and Yuyama, S. (1976). Cessation of nuclear DNA synthesis in differentiating *Naegleria*. J. Protozool. 23, 587-593.

Corff, S., and Yuyama, S. (1978). Rapid changes in the thymidine kinase activity of *Naegleria* induced to differentiate and revert under growth conditions. Exp. Cell Res. 114, 175-183.

Corff, S., Yuyama, S., and Young, P.G. (1982). Stimulation of thymidine kinase activity by chloramphenicol in *Naegleria*. J. Cell Physiol. 111, 33-41.

Culbertson, C.G., Ensminger, P.W., and Overton, W.M. (1968). Pathogenic *Naegleria* sp.--study of a strain isolated from human cerebrospinal fluid. J. Protozool. 15, 353-363.

Cursons, R.T., Brown, T.J., Keys, E.A., Gordon, E.H., Leng, R.H., Havill, J.H., and Hyne, B.E. (1979). Primary amoebic meningoencephalitis in an indoor heat-exchange swimming pool. N. Z. Med. J. 90, 330-331.

Cursons, R.T., Brown, T.J., and Keys, E.A. (1980). Effect of disinfectants on pathogenic free-living amoebae: in axenic conditions. Appl. Environ. Microbiol. 40, 62-66.

Cursons, R.T., Brown, T.J., Keys, E.A., Moriarty, K.M., and Till, D. (1980). Immunity to pathogenic free-living amoebae: role of humoral antibody. Infect. Immun. 29, 401-407.

da Silva, M.T., Caldas, V.E., Costa, F.C., Silvestre, D.A., and Thiemann, O.H. (2013). Selenocysteine biosynthesis and insertion machinery in *Naegleria gruberi*. Mol. Biochem. Parasitol. 188, 87-90.

Danso, S.K., and Alexander, M. (1975). Regulation of predation by prey density: the protozoan-*Rhizobium* relationship. Appl. Microbiol. 29, 515-521.

Das, S.R. (1974). Pathogenicity of flagellate stage of *Naegleria* aerobia and its bearing on the epidemiology of exogenous amoebiasis. Ann. Soc. Belg. Med. Trop. 54, 327-332.

Das, S., Saha, A.K., Nerad, T.A., Martinez, A.J., Lamarco, K.L., Basu, A., Legler, G., and Glew, R.H. 1987. Partial purification and characterization of *Naegleria fowleri* beta-glucosidase. J. Protozool. 34, 68-74.

da Rocha-Azevedo, B., Tanowitz, H.B., and Marciano-Cabral, F. (2009). Diagnosis of infections caused by pathogenic free-living amoebae. Interdiscip. Perspect. Infect. Dis. e251406.

da Silva, M.A., and da Rosa, J.A. (2003). Isolation of potencially pathogenic free-living amoebas in hospital dust. Rev. Saude Publica. 37, 242-246.

Debnath, A., Tunac, J.B., Galindo-Gómez, S., Silva-Olivares, A., Shibayama, M., and McKerrow, J.H. (2012). Corifungin, a new drug lead against *Naegleria*, identified from a high-throughput screen. Antimicrob. Agents Chemother. 56, 5450-5457.

Decatur, W.A., Johansen, S., and Vogt, V.M. (2000). Expression of the *Naegleria* intron endonuclease is dependent on a functional group I self-cleaving ribozyme. RNA 6, 616-627.

Declerck, P., Behets, J., Delaedt, Y., Margineanu, A., Lammertyn, E., and Ollevier, F. (2005). Impact of non-*Legionella* bacteria on the uptake and intracellular replication of *Legionella pneumophila* in *Acanthamoeba castellanii* and *Naegleria lovaniensis*. Microb. Ecol. 50, 536-549.

Declerck, P., Behets, J., De Keersmaecker, B., and Ollevier, F. (2007). Receptor-mediated uptake of *Legionella pneumophila* by *Acanthamoeba castellanii* and *Naegleria lovaniensis*. J. Appl. Microbiol. 103, 2697-2703.

De Jonckheere, J., Van Dijck, P., and Van de Voorde, H. (1974). Evaluation of the indirect fluorescent-antibody technique for identification of *Naegleria* species. Appl. Microbiol. 28, 159-164.

De Jonckheere, J., Van Dijck, P., and Van de Voorde, H. (1975). The effect of thermal pollution on the distribution of *Naegleria fowleri*. J. Hyg. (Lond). 75, 7-13.

De Jonckheere, J., and van de Voorde, H. (1976). Differences in destruction of cysts of pathogenic and nonpathogenic *Naegleria* and *Acanthamoeba* by chlorine. Appl. Environ. Microbiol. 31, 294-297.

De Jonckheere, J. (1977). Use of an axenic medium for differentiation between pathogenic and nonpathogenic *Naegleria fowleri* isolates. Appl. Environ. Microbiol. 33, 751-757.

De Jonckheere, J., and Voorde, H. (1977). The distribution of *Naegleria fowleri* in man-made thermal waters. Am. J. Trop. Med. Hyg. 26, 10-15.

De Jonckheere, J.F. (1979). Pathogenic free-living amoebae in swimming pools: survey in Belgium. Ann. Microbiol. (Paris). 130B, 205-212.

De Jonckheere, J. (1979). Differences in virulence of *Naegleria fowleri*. Pathol. Biol. (Paris). 27, 453-458.

De Jonckheere, J.F. (1982). Isoenzyme patterns of pathogenic and non-pathogenic *Naegleria* spp. using agarose isoelectric focusing. Ann. Microbiol. (Paris). 133, 319-342.

De Jonckheere, J.F. (1982). Hospital hydrotherapy pools treated with ultra violet light: bad bacteriological quality and presence of thermophilic *Naegleria*. J. Hyg. (Lond). 88, 205-214.

De Jonckheere, J.F., and Dierickx, P.J. (1982). Determination of acid phosphatase and leucine amino peptidase activity as an identification method for pathogenic *Naegleria fowleri*. Trans. R. Soc. Trop. Med. Hyg. 76, 773-775.

De Jonckheere, J.F., Aerts, M., and Martinez, A.J. (1983). *Naegleria australiensis*: experimental meningoencephalitis in mice. Trans. R. Soc. Trop. Med. Hyg. 77, 712-716.

De Jonckheere, J.F. (1987). Characterization of *Naegleria* species by restriction endonuclease digestion of whole-cell DNA. Mol. Biochem. Parasitol. 24, 55-66.

De Jonckheere, J.F. (1988). Geographic origin and spread of pathogenic *Naegleria fowleri* deduced from restriction

enzyme patterns of repeated DNA. Biosystems 21, 269-275.
De Jonckheere, J.F. (1989). Variation of electrophoretic karyotypes among *Naegleria* spp. Parasitol. Res. 76, 55-62.
De Jonckheere, J.F. (1993). A group I intron in the SSUrDNA of some *Naegleria* spp. demonstrated by polymerase chain reaction amplification. J. Eukaryot. Microbiol. 40, 179-187.
De Jonckheere, J.F. (1994). Riboprinting of *Naegleria* spp.: small-subunit versus large-subunit rDNA. Parasitol. Res. 80, 230-234.
De Jonckheere, J.F. (1994). Evidence for the ancestral origin of group I introns in the SSUrDNA of *Naegleria* spp. J. Eukaryot. Microbiol. 41, 457-463.
De Jonckheere, J.F., and Brown, S. (1994). Loss of the ORF in the SSUrDNA group I intron of one *Naegleria* lineage. Nucleic Acids Res. 22, 3925-3927.
De Jonckheere, J.F., and Brown, S. (1998). Three different group I introns in the nuclear large subunit ribosomal DNA of the amoeboflagellate *Naegleria*. Nucleic Acids Res. 26, 456-461.
De Jonckheere, J.F. (1998). Sequence variation in the ribosomal internal transcribed spacers, including the 5.8S rDNA, of *Naegleria* spp. Protist 149, 221-228.
De Jonckheere, J.F., Brown, S., Dobson, P.J., Robinson, B.S., and Pernin, P. (2001). The amoeba-to-flagellate transformation test is not reliable for the diagnosis of the genus *Naegleria*. Description of three new *Naegleria* spp. Protist 152, 115-121.
De Jonckheere, J.F. (2004). Molecular definition and the ubiquity of species in the genus *Naegleria*. Protist 155, 89-103.
De Jonckheere, J.F., and Brown, S. (2005). The identification of vahlkampfiid amoebae by ITS sequencing. Protist 156, 89-96.
De Jonckheere, J.F. (2006). Isolation and molecular identification of free-living amoebae of the genus *Naegleria* from Arctic and sub-Antarctic regions. Eur. J. Protistol. 42, 115-123.

De Jonckheere, J.F. (2011). Origin and evolution of the worldwide distributed pathogenic amoeboflagellate *Naegleria fowleri*. Infect. Genet. Evol. 11, 1520-1528.

Dhu, P. (1982). Effect of antimalarial drugs on *Naegleria fowleri*. Med. J. Aust. 1(1):13.

Dingle, A.D., and Fulton, C. (1966). Development of the flagellar apparatus of *Naegleria*. J. Cell Biol. 31, 43-54.

Dingle, A.D. (1970). Control of flagellum number in *Naegleria*. Temperature shock induction of multiflagellate cells. J. Cell Sci. 7, 463-481.

Dingle, A.D. (1979). Cellular and environmental variables determining numbers of flagella in temperature-shocked *Naegleria*. J. Protozool. 26, 604-612.

Dive, D.G., Leclerc, H., De Jonckheere, J., and Delattre, J.M. (1981). Isolation of *Naegleria fowleri* from the cooling pond of an electric power plant in France. Ann. Microbiol. (Paris). 132A, 97-105.

Dorsch, M.M., Cameron, A.S., and Robinson, B.S. (1983). The epidemiology and control of primary amoebic meningoencephalitis with particular reference to South Australia. Trans. R. Soc. Trop. Med. Hyg. 77, 372-377.

Dubray, B.L., Wilhelm, W.E., and Jennings, B.R. (1987). Serology of *Naegleria fowleri* and *Naegleria* lovaniensis in a hospital survey. J. Protozool. 34, 322-327.

Duma, R.J. (1970). *In vitro* susceptibility of pathogenic *Naegleria gruberi* to amphotericin B. Antimicrob. Agents Chemother. 10, 109-111.

Duma, R.J., Rosenblum, W.I., McGehee, R.F., Jones, M.M., and Nelson, E.C. (1971). Primary amoebic meningoencephalitis caused by *Naegleria*. Two new cases, response to amphotericin B, and a review. Ann. Intern. Med. 74, 923-931.

Duma, R.J., and Finley, R. (1976). *In vitro* susceptibility of pathogenic *Naegleria* and *Acanthamoeba* speicies to a variety of therapeutic agents. Antimicrob. Agents Chemother. 10, 370-376.

Dunnebacke, T.H., and Schuster, F.L. (1971). Infectious agent from a free-living soil amoeba, *Naegleria gruberi*. Science 174, 516-518.

Dunnebacke, T.H., and Schuster, F.L. (1974). An infectious agent associated with amoebas of the genus *Naegleria*. J. Protozool. 21, 327-329.

Dunnebacke, T.H., and Schuster, F.L. (1977). The nature of a cytopathogenic material present in amoebae of the genus *Naegleria*. Am. J. Trop. Med. Hyg. 26, 412-421.

Dunnebacke, T.H., and Schuster, F.L. (1985). Morphological response of cultured cells to *Naegleria* amoeba cytopathogenic material. J. Cell Sci. 75, 1-16.

Dunnebacke, T.H., and Dixon, J.S. (1990). NACM, a cytopathogenic protein from *Naegleria gruberi*, EGs; purification, production of monoclonal antibody, and the immunoidentification of a product that develops in NACM-treated vertebrate cell cultures. J. Protozool. 37, 11S-16S.

Dunnebacke, T.H., and Walen, K.H. (1999). A protein from *Naegleria* amoebae causes apoptosis in chick embryo and CHO cells after they become confluent. *In vitro* Cell. Dev. Biol. Anim. 35, 252-261.

Dyková, I., Kyselová, I., Pecková, H., Oborník, M., and Lukes, J. (2001). Identity of *Naegleria* strains isolated from organs of freshwater fishes. Dis. Aquat. Organ. 46, 115-121.

Einvik, C., Decatur, W.A., Embley, T.M., Vogt, V.M., and Johansen, S. (1997). *Naegleria* nucleolar introns contain two group I ribozymes with different functions in RNA splicing and processing. RNA 3, 710-20.

Einvik, C., Nielsen, H., Westhof, E., Michel, F., and Johansen, S. (1998). Group I-like ribozymes with a novel core organization perform obligate sequential hydrolytic cleavages at two processing sites. RNA 4, 530-541.

Eisen, D., and Franson, R.C. (1987). Acid-active neuraminidases in the growth media from cultures of pathogenic *Naegleria fowleri* and in sonicates of rabbit alveolar macrophages. Biochim. Biophys. Acta. 924, 369-372.

Elde, M., Haugen, P., Willassen, N.P., and Johansen, S. (1999). I-NjaI, a nuclear intron-encoded homing endonuclease from *Naegleria*, generates a pentanucleotide 3' cleavage-overhang within a 19 base-pair partially symmetric DNA recognition site. Eur. J. Biochem. 259, 281-288.

Elde, M., Willassen, N.P., and Johansen, S. (2000). Functional characterization of isoschizomeric His-Cys box homing endonucleases from *Naegleria*. Eur. J. Biochem. 267, 7257-7266.

Engel, J.P., Rubin, A.J., Sproul, O.J. (1983). Inactivation of *Naegleria gruberi* cysts by chlorinated cyanurates. Appl. Environ. Microbiol. 46, 1157-1162.

Ercken, D., Verelst, L., Declerck, P., Duvivier, L., Van Damme, A., and Ollevier, F. (2003). Effects of peracetic acid and monochloramine on the inactivation of *Naegleria lovaniensis*. Water Sci. Technol. 47, 167-171.

Esterman, A., Roder, D.M., Cameron, A.S., Robinson, B.S., Walters, R.P., Lake, J.A., and Christy, P.E. (1984). Determinants of the microbiological characteristics of South Australian swimming pools. Appl. Environ. Microbiol. 47, 325-328.

Feldman, M.R. (1977). *Naegleria fowleri*: fine structural localization of acid phosphatase and heme proteins. Exp. Parasitol. 41, 283-289.

Ferrante, A., and Thong, Y.H. (1979). Antibody induced capping and endocytosis of surface antigens in *Naegleria fowleri*. Int. J. Parasitol. 9, 599-601.

Ferrante, A. (1982). Comparative sensitivity of *Naegleria fowleri* to amphotericin B and amphotericin B methylester. Trans. R. Soc. Trop. Med. Hyg. 76, 476-478.

Ferrante, A., and Smyth, C. (1984). Mitogenicity of *Naegleria fowleri* extract for murine T lymphocytes. Immunol. 51, 461-468.

Ferrante, A., and Mocatta, T.J. (1984). Human neutrophils require activation by mononuclear leucocyte conditioned medium to kill the pathogenic free-living amoeba, *Naegleria fowleri*. Clin. Exp. Immunol. 56, 559-566.

Ferrante, A., and Lederer, E. (1986). Curative properties of muramyl dipeptide in experimental *Naegleria* meningo-encephalitis. Trans. R. Soc. Trop. Med. Hyg. 80, 323-326.

Ferrante, A., Hill, N.L., Abell, T.J., and Pruul, H. (1987). Role of myeloperoxidase in the killing of *Naegleria fowleri* by lymphokine-altered human neutrophils. Infect. Immun. 55, 1047-1050.

Ferrante, A., and Rowan-Kelly, B. (1988). The role of antibody in immunity against experimental *Naegleria* meningoencephalitis ('amoebic meningitis'). Immunol. 64, 241-244.

Ferrante, A., Carter, R.F., Lopez, A.F., Rowan-Kelly, B., Hill, N.L., and Vadas, M.A. (1988). Depression of immunity to *Naegleria fowleri* in mice by selective depletion of neutrophils with a monoclonal antibody. Infect. Immun. 56, 2286-2291.

Ferrante, A., and Bates, E.J. (1988). Elastase in the pathogenic free-living amoebae *Naegleria* and *Acanthamoeba* spp. Infect. Immun. 56, 3320-3321.

Ferrante, A. (1989). Augmentation of the neutrophil response to *Naegleria fowleri* by tumor necrosis factor alpha. Infect. Immun. 57, 3110-3115.

Ferrante, A., Hill, N.L., Goh, D.H., and Kumaratilake, L. (1989). Altered neutrophils in mice immune to experimental *Naegleria* amoebic meningoencephalitis. Immunol. Lett. 22, 301-305.

Fischer-Stenger, K., Cabral, G.A., and Marciano-Cabral, F. (1990). The interaction of *Naegleria fowleri* amoebae with murine macrophage cell lines. J. Protozool. 37, 168-173.

Fischer-Stenger, K., Cabral, G.A., and Marciano-Cabral, F. (1992). Separation of soluble amoebicidal and tumoricidal activity of activated macrophages. J. Protozool. 39, 235-241.

Fischer-Stenger, K., and Marciano-Cabral, F. (1992). The arginine-dependent cytolytic mechanism plays a role in destruction of *Naegleria fowleri* amoebae by activated macrophages. Infect. Immun. 60, 5126-5131.

Flores, B.M., Garcia, C.A., Stamm, W.E., and Torian, B.E. (1990). Differentiation of *Naegleria fowleri* from *Acanthamoeba* species by using monoclonal antibodies and flow cytometry. J. Clin. Microbiol. 28, 1999-2005.

Forrester, J.A., Gingell, D., and Korohoda, W. (1967). Electrophoretic polarity exhibited by amoeboid cells of *Naegleria gruberi*. Nature 215, 1409-1410.

Fowler, M., and Carter, R.F. (1965). Acute pyogenic meningitis probably due to *Acanthamoeba* sp.: a preliminary report. Br. Med. J. 2(5464), 740-742.

Fritz-Laylin, L.K., Assaf, Z.J., Chen, S., and Cande, W.Z. (2010). *Naegleria gruberi de novo* basal body assembly occurs via stepwise incorporation of conserved proteins. Eukaryot. Cell. 9, 860-865.

Fritz-Laylin, L.K., and Cande, W.Z. (2010). Ancestral centriole and flagella proteins identified by analysis of *Naegleria* differentiation. J. Cell Sci. 123, 4024-4031.

Fritz-Laylin, L.K., Ginger, M.L., Walsh, C., Dawson, S.C., and Fulton C. (2011). The *Naegleria* genome: a free-living microbial eukaryote lends unique insights into core eukaryotic cell biology. Res. Microbiol. 162, 607-618.

Fritzinger, A.E., and Marciano-Cabral, F. (2004). Modulation of a "CD59-like" protein in *Naegleria fowleri* amoebae by bacteria. J. Eukaryot. Microbiol. 51, 522-528.

Fritzinger, A.E., Toney, D.M., MacLean, R.C., and Marciano-Cabral, F. (2006). Identification of a *Naegleria fowleri* membrane protein reactive with anti-human CD59 antibody. Infect. Immun. 74, 1189-1195.

Fulford, D.E., Bradley, S.G., and Marciano-Cabral, F. (1984). Cytopathogenicity of *Naegleria fowleri* for cultured rat neuroblastoma cells. J. Protozool. 32, 176-180.

Fulford, D.E., and Marciano-Cabral, F. (1986). Cytolytic activity of *Naegleria fowleri* cell-free extract. J. Protozool. 33, 498-502.

Fulton, C., and Dingle, A.D. (1967). Appearance of the flagellate phenotype in populations of *Naegleria* amoebae. Dev. Biol. 15, 165-191.

Fulton, C., and Dingle, A.D. (1971). Basal bodies, but not centrioles, in *Naegleria*. J. Cell Biol. 51, 826-836.

Fulton, C. (1974). Axenic cultivation of *Naegleria gruberi*. Requirement for methionine. Exp. Cell Res. 88, 365-370.

Fulton, C., and Kowit, J.D. (1975). Programmed synthesis of flagellar tubulin during cell differentiation in *Naegleria*. Ann. N. Y. Acad. Sci. 253, 318-332.

Fulton, C. (1977). Intracellular regulation of cell shape and motility in *Naegleria*. First insights and a working hypothesis. J. Supramol. Struct. 6, 13-43.

Fulton, C., and Walsh, C. (1980). Cell differentiation and flagellar elongation in *Naegleria gruberi*. Dependence on transcription and translation. J. Cell Biol. 85, 346-360.

Fulton, C., Webster, C., and Wu, J.S. (1984). Chemically defined media for cultivation of *Naegleria gruberi*. Proc. Natl. Acad. Sci. USA. 81, 2406-2410.

Fulton, C., Cheng, K.L., and Lai, E.Y. (1986). Two calmodulins in *Naegleria* flagellates: characterization, intracellular segregation, and programmed regulation of mRNA abundance during differentiation. J. Cell Biol. 102, 1671-1678.

Fulton, C., Lai, E.Y., Lamoyi, E., and Sussman, D.J. (1986). *Naegleria* actin elicits species-specific antibodies. J. Protozool. 33, 322-327.

Fulton, C., Lai, E.Y., and Remillard, S.P. (1995). A flagellar calmodulin gene of *Naegleria*, coexpressed during differentiation with flagellar tubulin genes, shares DNA, RNA, and encoded protein sequence elements. J. Biol. Chem. 270, 5839-5848.

Fulton, C., and Lai, E.Y. (1998). Stable intermediates and holdpoints in the rapid differentiation of *Naegleria*. Exp. Cell Res. 242, 429-438.

Futch, W.S.Jr., and Schook, L.B. (1985). Dissection of macrophage tumoricidal and protozoacidal activities using T-cell hybridomas and recombinant lymphokines. Infect. Immun. 50, 709-715.

Gálvez, A., Valdivia, E., González-Segura, A., Lebbadi, M., Martínez-Bueno, M., and Maqueda, M. (1993). Purification, characterization, and lytic activity against *Naegleria fowleri* of two amoebicins produced by *Bacillus licheniformis* A12. Appl. Environ. Microbiol. 59, 1480-1486.

Gálvez, A., Maqueda, M., Cordovilla, P., Martínez-Bueno, M., Lebbadi, M., and Valdivia, E. (1994). Characterization and biological activity against *Naegleria fowleri* of amoebicins produced by *Bacillus licheniformis* D-13. Antimicrob. Agents Chemother. 38, 1314-1319.

Gardiner, P.R., Miller, R.H., and Marsh, M.C. (1981). Studies of the rhizoplast from *Naegleria gruberi*. J. Cell Sci. 47, 277-293.

Gianinazzi, C., Schild, M., Müller, N., Leib, S.L., Simon, F., Nuñez, S., Joss, P., and Gottstein, B. (2005). Organotypic slice cultures from rat brain tissue: a new approach for *Naegleria fowleri* CNS infection *in vitro*. Parasitol. 131, 797-804.

Ginger, M.L., Fritz-Laylin, L.K., Fulton, C., Cande, W.Z., and Dawson, S.C. (2010). Intermediary metabolism in protists: a sequence-based view of facultative anaerobic metabolism in evolutionarily diverse eukaryotes. Protist 161, 642-671.

Gogate, A., and Dadkar, V. (1986). Effect of some antimicrobial agents in primary amoebic meningo-encephalitis in mice. Indian J. Med. Res. 83, 148-151.

González-Robles, A., Castañón, G., Cristóbal-Ramos, A.R., Hernández-Ramírez, V.I., Omaña-Molina, M., and Martínez-Palomo, A. (2007). Cell surface differences of *Naegleria fowleri* and *Naegleria lovaniensis* exposed with surface markers. Exp. Parasitol. 117, 399-404.

González-Robles, A., Cristóbal-Ramos, A.R., González-Lázaro, M., Omaña-Molina, M., and Martínez-Palomo, A. (2009). *Naegleria fowleri*: light and electron microscopy study of mitosis. Exp. Parasitol. 122, 212-217.

Gorospe, S., Band, R.N., and Kopachik, W.J. (1996). Molecular cloning of, and phylogenetic analysis of, an actin in *Naegleria fowleri*. FEMS Microbiol. Lett. 141, 233-237.

Goswick, S.M., and Brenner, G.M. (2003). Activities of azithromycin and amphotericin B against *Naegleria fowleri in vitro* and in a mouse model of primary amoebic meningoencephalitis. Antimicrob. Agents Chemother. 47, 524-528.

Goswick, S.M., and Brenner, G.M. (2003). Activities of therapeutic agents against *Naegleria fowleri in vitro* and in a mouse model of primary amoebic meningoencephalitis. J. Parasitol. 89, 837-842.

Goudot, S., Herbelin, P., Mathieu, L., Soreau, S., Banas, S., and Jorand, F. (2012). Growth dynamic of *Naegleria fowleri* in a microbial freshwater biofilm. Water Res. 46, 3958-3966.

Goudot, S., Herbelin, P., Mathieu, L., Soreau, S., Banas, S., and Jorand, F.P. (2014). Biocidal efficacy of monochloramine against planktonic and biofilm-associated *Naegleria fowleri* cells. J. Appl. Microbiol. 116, 1055-1065.

Gray, S.A., Weigel, K.M., Ali, I.K., Lakey, A.A., Capalungan, J., Domingo, G.J., and Cangelosi, G.A. (2012). Toward low-cost affinity reagents: lyophilized yeast-scFv probes specific for pathogen antigens. PLoS One **7,** e32042.

Grewal, Y.S., Shiddiky, M.J., Gray, S.A., Weigel, K.M., Cangelosi, G.A., and Trau, M. (2013). Label-free electrochemical detection of an *Entamoeba histolytica* antigen using cell-free yeast-scFv probes. Chem. Commun. (Camb) 49, 1551-1553.

Grewal, Y.S., Shiddiky, M.J., Spadafora, L.J., Cangelosi, G.A., and Trau, M. (2014). Nano-yeast-scFv probes on screen-printed gold electrodes for detection of *Entamoeba histolytica* antigens in a biological matrix. Biosens. Bioelectron 55, 417-422.

Griffin, J.L. (1972). Temperature tolerance of pathogenic and nonpathogenic free-living amoebas. Science 178, 869-870.

Griffin, J.L. (1983). The pathogenic amoeboflagellate *Naegleria fowleri*: environmental isolations, competitors, ecologic interactions, and the flagellate-empty habitat hypothesis. J. Protozool. 30, 403-409.

Grimm, D., Ludwig, W.F., Brandt, B.C., Michel, R., Schleifer, K.H., Hacker, J., Steinert, M. (2001). Development of 18S rRNA-targeted oligonucleotide probes for specific detection of *Hartmannella* and *Naegleria* in *Legionella*-positive environmental samples. Syst. Appl. Microbiol. 24, 76-82.

Guarner, J., Bartlett, J., Shieh, W.J., Paddock, C.D., Visvesvara, G.S., and Zaki, S.R. (2007). Histopathologic spectrum and immunohistochemical diagnosis of amoebic meningoencephalitis. Mod. Pathol. 20, 1230-1237.

Gupta, S., Ghosh, P.K., Dutta, G.P., and Vishwakarma, R.A. (1995). *In vivo* study of artemisinin and its derivatives against primary amoebic meningoencephalitis caused by *Naegleria fowleri*. J. Parasitol. 81, 1012-1013.

Gupta, S., Dutta, G.P., and Vishwakarma, R.A. (1998). Effect of alpha,beta-arteether against primary amoebic meningoencephalitis in Swiss mice. Indian J. Exp. Biol. 36, 824-825.

Gupta, S., and Das, S.R. (1999). Stock cultures of free-living amoebas: effect of temperature on viability and pathogenicity. J. Parasitol. 85, 137-139.

Gustavo dos Santos, N. (1970). Fatal primary amoebic meningoencephalitis. A retrospective study in Richmond, Virginia. Am. J. Clin. Pathol. 54, 737-742.

Haggerty, R.M., and John, D.T. (1978). Innate resistance of mice to experimental infection with *Naegleria fowleri*. Infect. Immun. 20, 73-77.

Haggerty, R.M., and John, D.T. (1982). Serum agglutination and immunoglobulin levels of mice infected with *Naegleria fowleri*. J. Protozool. 29, 117-122.

Haight, J.B., and John, D.T. (1980). Growth of *Naegleria fowleri* in several axenic media. Folia Parasitol. (Praha). 27, 207-212.

Hambidge, A. (2001). Reviewing efficacy of alternative water treatment techniques. Health Estate. 55, 23-25.

Han, K.H., Ahn, M.H., and Min, D.Y. (1988). Immunization effect on *Naegleria fowleri* infection in splenectomized mice. Kisaengchunghak Chapchi. 26, 39-44.

Han, J.W., Park, J.H., Kim, M., and Lee, J. (1997). mRNAs for microtubule proteins are specifically colocalized during the sequential formation of basal body, flagella, and cytoskeletal microtubules in the differentiation of *Naegleria gruberi*. J. Cell Biol. 137, 871-879.

Han, K.L., Lee, H.J., Shin, M.H., Shin, H.J., Im, K.I., and Park, S.J. (2004). The involvement of an integrin-like protein and protein kinase C in amoebic adhesion to fibronectin and amoebic cytotoxicity. Parasitol. Res. 94, 53-60.

Hashimoto, H., Pais, J.E., Zhang, X., Saleh, L., Fu, Z.Q., Dai, N., Corrêa, I.R. Jr., Zheng, Y., and Cheng, X. (2014). Structure of a *Naegleria* Tet-like dioxygenase in complex with 5-methylcytosine DNA. Nature 506(7488), 391-395.

Hara, T., and Fukuma, T. (2005). Diagnosis of the primary amoebic meningoencephalitis due to *Naegleria fowleri*. Parasitol. Int. 54, 219-221.

Haugen, P., De Jonckheere, J.F., and Johansen, S. (2002). Characterization of the self-splicing products of two complex *Naegleria* LSU rDNA group I introns containing homing endonuclease genes. Eur. J. Biochem. 269, 1641-1649.

Heaselgrave, W., and Kilvington, S. (2011). The efficacy of simulated solar disinfection (SODIS) against *Ascaris*, *Giardia*, *Acanthamoeba*, *Naegleria*, *Entamoeba* and *Cryptosporidium*. Acta Trop. 119, 138-143.

Herbst, R., Ott, C., Jacobs, T., Marti, T., Marciano-Cabral, F., and Leippe, M. (2002). Pore-forming polypeptides of the pathogenic protozoon *Naegleria fowleri*. J. Biol. Chem. 277, 22353-22360.

Herbst, R., Marciano-Cabral, F., and Leippe, M. (2004). Antimicrobial and pore-forming peptides of free-living and potentially highly pathogenic *Naegleria fowleri* are released from the same precursor molecule. J. Biol. Chem. 279, 25955-25958.

Herman, E.K., Greninger, A.L., Visvesvara, G.S., Marciano-Cabral, F., Dacks, J.B., and Chiu, C.Y. (2013). The mitochondrial genome and a 60-kb nuclear DNA segment from *Naegleria fowleri*, the causative agent of primary amoebic meningoencephalitis. J. Eukaryot. Microbiol. 60, 179-191.

Hickey, C., Nakazato, H., Edmonds, M., and Walsh, C. (1979). Poly(a) sequences in *Naegleria* messenger RNA before and after initiation of differentiation. Absence of oligo(A). Exp. Cell Res. 120, 277-284.

Hillebrandt, S., and Müller, I. (1991). Repair of damage caused by UV- and X-irradiation in the amoeboflagellate *Naegleria gruberi*. Radiat. Environ. Biophys. 30, 123-130.

Hobson, R.P. (2000). The effects of diffusates from the spores of *Aspergillus fumigatus* and *A. terreus* on human neutrophils, *Naegleria gruberi* and *Acanthamoeba castellanii*. Med. Mycol. 38, 133-141.

Holbrook, T.W., and Parker, B.W. (1979). *Naegleria fowleri* in chick embryos. Effects of embryo age and incubation temperature, and the infectivity of embryo-derived amoebae for mice. Am. J. Trop. Med. Hyg. 28, 984-987.

Holbrook, T.W., Boackle, R.J., Parker, B.W., and Vesely, J. (1980). Activation of the alternative complement pathway by *Naegleria fowleri*. Infect. Immun. 30, 58-61.

Hu, W.N., Band, R.N., and Kopachik, W.J. (1991). Virulence-related protein synthesis in *Naegleria fowleri*. Infect. Immun. 59, 4278-4282.

Hu, W.N., Kopachik, W., and Band, R.N. (1992). Cloning and characterization of transcripts showing virulence-related gene expression in *Naegleria fowleri*. Infect. Immun. 60, 2418-2424.

Huang, S.W., Hsu, B.M., Chen, N.H., Huang, C.C., Huang, K.H., Chen, J.S., and Kao, P.M. (2011). Isolation and identification of *Legionella* and their host amoebae from weak alkaline carbonate spring water using a culture method combined with PCR. Parasitol. Res. 109, 1233-1241.

Hysmith, R.M., and Franson, R.C. (1982). Degradation of human myelin phospholipids by phospholipase-enriched culture media of pathogenic *Naegleria fowleri*. Biochim. Biophys. Acta. 712, 698-701.

Hysmith, R.M., and Franson, R.C. (1982). Elevated levels of cellular and extracellular phospholipases from pathogenic *Naegleria fowleri*. Biochim. Biophys. Acta. 711, 26-32.

Im, K.I., and Lee, K.T. (1985). Failure of immunization with *Naegleria fowleri* in mice born to immune mothers. Kisaengchunghak Chapchi. 23, 151-155.

Im, K.I., Ryu, J.S., and Lee, K.T. (1987). Immunodepression during experimental *Naegleria* meningoencephalitis in mice. Kisaengchunghak Chapchi. 25, 195-198.

Im, K.I., and Ryu, J.S. (1988). Passive immunity by splenocyte transfer against amoebic meningoencephalitis in mice. Kisaengchunghak Chapchi. 26, 169-173.

Jabri, E., Aigner, S., and Cech, T.R. (1997). Kinetic and secondary structure analysis of *Naegleria andersoni* GIR1, a group I ribozyme whose putative biological function is site-specific hydrolysis. Biochem. 36, 16345-16354.

Jabri, E., and Cech, T.R. (1998). *In vitro* selection of the *Naegleria* GIR1 ribozyme identifies three base changes that dramatically improve activity. RNA 4, 1481-1492.

Jamerson, M., da Rocha-Azevedo, B., Cabral, G.A., and Marciano-Cabral, F. (2012). Pathogenic *Naegleria fowleri* and non-pathogenic *Naegleria lovaniensis* exhibit

differential adhesion to, and invasion of, extracellular matrix proteins. Microbiol. 158, 791-803.
Jamieson, A., and Anderson, K. (1973). A method for the isolation of *Naegleria* species from water samples. Pathol. 5, 55-58.
Jamieson, A. (1975). Effect of clotrimazole on *Naegleria fowleri*. J. Clin. Pathol. 28, 446-449.
Jarillo-Luna, A., Moreno-Fierros, L., Campos-Rodríguez, R., Rodríguez-Monroy, M.A., Lara-Padilla, E., and Rojas-Hernández, S. (2007). Intranasal immunization with *Naegleria fowleri* lysates and Cry1Ac induces metaplasia in the olfactory epithelium and increases IgA secretion. Parasite Immunol. 30, 31-38.
Jaroli, K.L., McCosh, J.K., and Howard, M.J. (2002). The role of blood vessels and lungs in the dissemination of *Naegleria fowleri* following intranasal inoculation in mice. Folia Parasitol. (Praha). 49, 183-188.
Jarolim, K.L., McCosh, J.K., Howard, M.J., and John, D.T. (2000). A light microscopy study of the migration of *Naegleria fowleri* from the nasal submucosa to the central nervous system during the early stage of primary amoebic meningoencephalitis in mice. J. Parasitol. 86, 50-55.
Jeffery, S., and Hawkins, S.E. (1975). Studies on transformation in *Naegleria gruberi*: effects of ions and bacterial suspensions. Microbios. 15, 27-36.
Jeong, S.R., Kang, S.Y., Lee, S.C., Song, K.J., Im, K.I., and Shin, H.J. (2004). Decreasing effect of an anti-Nfa1 polyclonal antibody on the *in vitro* cytotoxicity of pathogenic *Naegleria fowleri*. Kor. J. Parasitol. 42, 35-40.
Jeong, S.R., Lee, S.C., Song, K.J., Park, S., Kim, K., Kwon, M.H., Im, K.I., and Shin, H.J. (2005). Expression of the nfa1 gene cloned from pathogenic *Naegleria fowleri* in nonpathogenic *N. gruberi* enhances cytotoxicity against CHO target cells *in vitro*. Infect. Immun. 73, 4098-4105.
John, D.T., Weik, R.R., and Adams, A.C. (1977). Immunization of mice against *Naegleria fowleri* infection. Infect. Immun. 16, 817-820.
John, D.T. (1982). Primary amoebic meningoencephalitis and the biology of *Naegleria fowleri*. Ann. Rev. Microbiol. 36, 101-123.

John, D.T., and Nussbaum, S.L. (1983). *Naegleria fowleri* infection acquired by mice through swimming in amoebae-contaminated water. J. Parasitol. 69, 871-874.

John, D.T., Cole, T.B.Jr., and Marciano-Cabral, F.M. (1984). Sucker-like structures on the pathogenic amoeba *Naegleria fowleri*. Appl. Environ. Microbiol. 47, 12-14.

John, D.T., Cole,T.B.Jr., and Bruner, R.A. (1984). Amoebostomes of *Naegleria fowleri*. J. Protozool. 32, 12-19.

John, D.T., and John, R.A. (1988). Cytopathogenicity of *Naegleria fowleri* in mammalian cell cultures. Parasitol. Res. 76, 20-25.

John, D.T., and Hoppe, K.L. (1990). Susceptibility of wild mammals to infection with *Naegleria fowleri*. J. Parasitol. 76, 865-868.

John, D.T., Cole, T.B.Jr., and John, R.A. (1991). Flagella number among *Naegleria* flagellates. Folia Parasitol. (Praha). 38, 289-295.

John, D.T., and Howard, M.J. (1993). Virulence of *Naegleria fowleri* affected by axenic cultivation and passage in mice. Folia Parasitol. (Praha). 40, 187-191.

John, D.T., and John, R.A. (1994). Enhancement of virulence of *Naegleria fowleri* by growth in Vero-cell cultures. J. Parasitol. 80, 149-151.

John, D.T., Eddy, P.L., and John, R.A. (1994). Cryopreservation of pathogenic free-living amoebae. Folia Parasitol. (Praha). 41, 110-114.

John, D.T., and McCutchen, C.V. (1995). Reduction in virulence of *Naegleria fowleri* following growth with cholesterol. Folia Parasitol. (Praha). 42, 236-238.

John, D.T., and John, R.A. (1996). Viability of pathogenic *Acanthamoeba* and *Naegleria* and virulence of *N. fowleri* during long-term cryopreservation. Folia Parasitol. (Praha). 43, 43-46.

John, D.T., and Smith, B.L. (1997). Amoebicidal activity of wild animal serum. J. Parasitol. 83, 757-759.

John, D.T., and John, R.A. (2006). Viability of pathogenic *Naegleria* and *Acanthamoeba* isolates during 10 years of cryopreservation. Folia Parasitol. (Praha). 53, 311-312.

Johnson, P.E., Deromedi, A.J., Lebaron, P., Catala, P., Havens, C., and Pougnard, C. (2007). High throughput, real-time detection of *Naegleria lovaniensis* in natural river water using LED-illuminated Fountain Flow Cytometry. J. Appl. Microbiol. 103, 700-710.

Josephson, S.L., Weik, R.R., and John, D.T. (1977). Concanavalin A-induced agglutination of *Naegleria*. Am. J. Trop. Med. Hyg. 26, 856-858.

Jung, S.Y., Kim, J.H., Lee, Y.J., Song, K.J., Kim, K., Park, S., Im, K.I., and Shin, H.J. (2007). *Naegleria fowleri*: *nfa1* gene knock-down by double-stranded RNAs. Exp. Parasitol. 118, 208-213.

Jung, S.Y., Kim, J.H., Song, K.J., Lee, Y.J., Kwon, M.H., Kim, K., Park, S., Im, K.I., and Shin, H.J. (2009). Gene silencing of nfa1 affects the *in vitro* cytotoxicity of *Naegleria fowleri* in murine macrophages. Mol. Biochem. Parasitol. 165, 87-93.

Kadlec, V., Cerva, L., and Skvárová, J. (1978). Virulent *Naegleria fowleri* in an indoor swimming pool. Science 201(4360):1025.

Kang, S.Y., Song, K.J., Jeong, S.R., Kim, J.H., Park, S., Kim, K., Kwon, M.H., and Shin, H.J. (2005). Role of the Nfa1 protein in pathogenic *Naegleria fowleri* cocultured with CHO target cells. Clin. Diagn. Lab. Immunol. 12, 873-876.

Kasprzak, W., Mazur, T., and Cerva, L. (1982). *Naegleria fowleri* in thermally polluted waters). Folia Parasitol. (Praha). 29, 211-218.

Kao, P.M., Hsu, B.M., Hsu, T.K., Chiu, Y.C., Chang, C.L., Ji, W.T., Huang, S.W., and Fan, C.W. (2014). Application of TaqMan qPCR for the detection and monitoring of *Naegleria* species in reservoirs used as a source for drinking water. Parasitol. Res. 113, 3765-3771.

Kilvington, S., and White, D.G. (1985). Rapid identification of thermophilic *Naegleria*, including *Naegleria fowleri* using API ZYM system. J. Clin. Pathol. 38, 1289-1292.

Kilvington, S., and White, D.G. (1986). Identification of *Naegleria fowleri* in fresh isolates of environmental amoebae using a staphylococcal coagglutination technique. Trans. R. Soc. Trop. Med. Hyg. 80, 564-569.

Kilvington, S., and White, D. (1991). A simple method for the cryopreservation of free-living amoebae belonging to the genera *Naegleria* and *Acanthamoeba*. Eur. J. Protistol. 27, 115-118.

Kilvington, S., and Beeching, J. (1995). Development of a PCR for identification of *Naegleria fowleri* from the environment. Appl. Environ. Microbiol. 61, 3764-3767.

Kim, O.Y., Im, K.I., Lee, K.T., and Choe, R.S. (1985). Effect of prednisolone treatment on the experimental inducement of primary amoebic meningoencephalitis. Kisaengchunghak Chapchi. 23, 165-172.

Kim, K.H., Shin, C.O., and Im, K. (1993). Natural killer cell activity in mice infected with free-living amoeba with reference to their pathogenicity. Kor. J. Parasitol. 31, 239-248.

Kim, H.K., Kang, J.G., Yumura, S., Walsh, C.J., Cho, J.W., and Lee, J. (2005). *De novo* formation of basal bodies in *Naegleria gruberi*: regulation by phosphorylation. J. Cell Biol. 169, 719-724.

Kim, J.H., Lee, Y.J., Sohn, H.J., Song, K.J., Kwon, D., Kwon, M.H., Im, K.I., and Shin, H.J. (2008). Therapeutic effect of rokitamycin *in vitro* and on experimental meningoencephalitis due to *Naegleria fowleri*. Int. J. Antimicrob. Agents 32, 411-417.

Kim, J.H., Jung, S.Y., Lee, Y.J., Song, K.J., Kwon, D., Kim, K., Park, S., Im, K.I., and Shin, H.J. (2008). Effect of therapeutic chemical agents *in vitro* and on experimental meningoencephalitis due to *Naegleria fowleri*. Antimicrob. Agents Chemother. 52, 4010-4016.

Kim, J.H., Kim, D., and Shin, H.J. (2008). Contact-independent cell death of human microglial cells due to pathogenic *Naegleria fowleri* trophozoites. Kor. J. Parasitol. 46, 217-221.

Kim, J.H., Yang, A.H., Sohn, H.J., Kim, D., Song, K.J., and Shin, H.J. (2009). Immunodominant antigens in *Naegleria fowleri* excretory--secretory proteins were potential pathogenic factors. Parasitol. Res. 105, 1675-1681.

Kim, J.H., Sohn, H.J., Lee, S.H., Kwon, D., and Shin, H.J. (2012). Induction of interleukin-8 by *Naegleria fowleri* lysates requires activation of extracellular signal-regulated kinase in human astroglial cells. Parasitol. Res. 111, 587-592.

Kim, J.H., Lee, S.H., Sohn, H.J., Lee, J., Chwae, Y.J., Park, S., Kim, K., and Shin, H.J. (2012). The immune response induced by DNA vaccine expressing *nfa1* gene against *Naegleria fowleri*. Parasitol. Res. 111, 2377-2384.

Kim, J.H., Song, A.R., Sohn, H.J., Lee, J., Yoo, J.K., Kwon, D., and Shin, H.J. (2013). IL-1β and IL-6 activate inflammatory responses of astrocytes against *Naegleria fowleri* infection via the modulation of MAPKs and AP-1. Parasite Immunol. 35, 120-128.

Kim, J.H., Sohn, H.J., Lee, J., Yang, H.J., Chwae, Y.J., Kim, K., Park, S., and Shin, H.J. (2013). Vaccination with lentiviral vector expressing the nfa1 gene confers a protective immune response to mice infected with *Naegleria fowleri*. Clin. Vaccine Immunol. 20, 1055-1060.

King, C.A., and Preston, T.M. (1977). Studies of anionic sites on the cell surface of the amoeba *Naegleria gruberi* using cationized ferritin. J. Cell Sci. 28, 133-149.

King, C.A., Preston, T.M., and Miller, R.H. (1983). Cell-substrate interactions in amoeboid locomotion - a matched reflexion interference and transmission electron microscopy study. Cell Biol. Int. Rep. 7, 641-649.

Knoop, V., and Rüdinger, M. (2010). DYW-type PPR proteins in a heterolobosean protist: plant RNA editing factors involved in an ancient horizontal gene transfer? FEBS Lett. 584, 4287-4291.

Kodousek, R., and Schrottenbaum, M. (1982). Penetration of amoebae into ganglion cells of the cerebral cortex in primary amoebic meningoencephalitis. Cesk Patol. 18, 50-54.

Kollars, T.M.Jr., and Wilhelm, W.E. (1996). The occurrence of antibodies to *Naegleria* species in wild mammals. J. Parasitol. 82, 73-77.

Kowit, J.D., and Fulton, C. (1974). Purification and properties of flagellar outer doublet tubulin from *Naegleria gruberi* and a radioimmune assay for tubulin. J. Biol. Chem. 249, 3638-3646.

Kowit, J.D., and Fulton, C. (1974). Programmed synthesis of tubulin for the flagella that develop during cell differentiation in *Naegleria gruberi*. Proc. Natl. Acad. Sci. USA. 71, 2877-2881.

Kreinbring, C.A., Remillard, S.P., Hubbard, P., Brodkin, H.R., Leeper, F.J., Hawksley, D., Lai, E.Y., Fulton, C., Petsko, G.A., and Ringe, D. (2014). Structure of a eukaryotic thiaminase I. Proc. Natl. Acad. Sci. USA. 111, 137-142.

Kyle, D.E., and Noblet, G.P. (1986). Seasonal distribution of thermotolerant free-living amoebae. I. Willard's Pond. J. Protozool. 33, 422-434.

Kyle, D.E., and Noblet, G.P. (1987). Seasonal distribution of thermotolerant free-living amoebae. II. Lake Issaqueena. J. Protozool. 34, 10-15.

Lai, E.Y., Walsh, C., Wardell, D., and Fulton, C. (1979). Programmed appearance of translatable flagellar tubulin mRNA during cell differentiation in *Naegleria*. Cell 17, 867-878.

Lai, E.Y., Remillard, S.P., and Fulton, C. (1988). The alpha-tubulin gene family expressed during cell differentiation in *Naegleria gruberi*. J. Cell Biol. 106, 2035-2046.

Lai, E., Remillard, S.P., and Fulton, C. (1994). A beta-tubulin gene of *Naegleria* encodes a carboxy-terminal tyrosine. Aromatic amino acids are conserved at carboxy termini. J. Mol. Biol. 235, 377-388.

Lallinger, G.J., Reiner, S.L., Cooke, D.W., Toffaletti, D.L., Perfect, J.R., Granger, D.L., and Durack, D.T. (1987). Efficacy of immune therapy in early experimental *Naegleria fowleri* meningitis. Infect. Immun. 55, 1289-1293.

Lam, A.H., de Silva, M., Procopis, P., and Kan, A. (1982). Primary amoebic (*Naegleria*) meningoencephalitis. J. Comput. Assist. Tomogr. 6, 620-623.

Larson, D.E., and Dingle, A.D. (1981). Isolation, ultrastructure, and protein composition of the flagellar rootlet of *Naegleria gruberi*. J. Cell Biol. 89, 424-432.

Larson, D.E., and Dingle, A.D. (1981). Development of the flagellar rootlet during *Naegleria* flagellate differentiation. Dev. Biol. 86, 227-235.

Lastovica, A.J., and Dingle, A.D. (1971). Superprecipitation of an actomyosin-like complex isolated from *Naegleria gruberi* amoebae. Exp. Cell Res. 66, 337-345.

Lastovica, A.J. (1974). Scanning electron microscopy of pathogenic and non-pathogenic *Naegleria* cysts. Int. J. Parasitol. 4, 139-142.

Lastovica, A.J. (1976). Microfilaments in *Naegleria fowleri* amoebae. Z. Parasitenkd. 50, 245-250.

Lawande, R.V., Abraham, S.N., John, I., and Egler, L.J. (1979). Recovery of soil Amoebas from the nasal passages of children during the dusty harmattan period in Zaria. Am. J. Clin. Pathol. 71, 201-203.

Lawande, R.V. (1983). Recovery of soil amoebae from the air during the harmattan in Zaria, Nigeria. Ann. Trop. Med. Parasitol. 77, 45-49.

Lebbadi, M., Gálvez, A., Valdivia, E., Martínez-Bueno, M., and Maqueda, M. (1994). Purification of amoebolytic substances from *Bacillus licheniformis* M-4. Arch. Microbiol. 162, 98-102.

Lebbadi, M., Gálvez, A., Valdivia, E., Martínez-Bueno, M., Maqueda, M. (1994). Biological activity of amoebicin m4-A from *Bacillus licheniformis* M-4. Antimicrob. Agents Chemother. 38, 1820-1823.

Lebbadi, M., Valdivia, E., Gálvez, A., Martínez-Bueno, M., and Maqueda, M. (1995). Cocultivation of the amoeba *Naegleria fowleri* and the amoebicin- producing strain *Bacillus licheniformis* M-4. Appl. Environ. Microbiol. 61, 1649-1652.

Le Calvez, T., Trouilhé, M.C., Humeau, P., Moletta-Denat, M., Frère, J., and Héchard, Y. (2012). Detection of free-living amoebae by using multiplex quantitative PCR. Mol. Cell. Probes 26, 116-120.

Lee, K.K., Karr, S.L.Jr., Wong, M.M., and Hoeprich, P.D. (1979). *In vitro* susceptibilities of *Naegleria fowleri* strain HB-1 to selected antimicrobial agents, singly and in combination. Antimicrob. Agents Chemother. 16, 217-220.

Lee, D.K., Lee, K.T., and Im, K.I. (1983). Changes in the pathogenicity of *Naegleria fowleri* by several brain passage in mice. Kisaengchunghak Chapchi. 21, 234-240.

Lee, S.G., Im, K.I., and Lee, K.T. (1985). Protective immunity against *Naegleria* meningoencephalitis in mice. Kisaengchunghak Chapchi. 23, 293-299.

Lee, J.H., and Walsh, C.J. (1988). Transcriptional regulation of coordinate changes in flagellar mRNAs during differentiation of *Naegleria gruberi* amoebae into flagellates. Mol. Cell. Biol. 8, 2280-2287.

Lee, S.G., Shin, H.J., and Im, K.I. (1989). Studies on the cell-mediated immunity in experimental *Naegleria* spp. infections. Kisaengchunghak Chapchi. 27, 177-186.

Lee, K.R., Shin, C.O., and Im, K. (1991). Natural killer cell activity in *Naegleria fowleri* infected mice. Kisaengchunghak Chapchi. 29, 267-277.

Lee, Y.J., Kim, J.H., Jeong, S.R., Song, K.J., Kim, K., Park, S., Park, M.S., and Shin, H.J. (2007). Production of Nfa1-specific monoclonal antibodies that influences the *in vitro* cytotoxicity of *Naegleria fowleri* trophozoites on microglial cells. Parasitol. Res. 101, 1191-1196.

Lee, J. (2009). *De novo* formation of basal bodies during cellular differentiation of *Naegleria gruberi*: progress and hypotheses. Semin. Cell Dev. Biol. 21, 156-162.

Lee, Y.J., Kim, J.H., Sohn, H.J., Lee, J., Jung, S.Y., Chwae, Y.J., Kim, K., Park, S., and Shin, H.J. (2011). Effects of immunization with the rNfa1 protein on experimental *Naegleria fowleri*-PAM mice. Parasite Immunol. 33, 382-389.

Lee, Y.J., Park, C.E., Kim, J.H., Sohn, H.J., Lee, J., Jung, S.Y., and Shin, H.J. (2011). *Naegleria fowleri* lysate induces strong cytopathic effects and pro-inflammatory cytokine release in rat microglial cells. Kor. J. Parasitol. 49, 285-290.

Lee, J., Kim, J.H., Sohn, H.J., Yang, H.J., Na, B.K., Chwae, Y.J., Park, S., Kim, K., and Shin, H.J. (2014). Novel cathepsin B and cathepsin B-like cysteine protease of *Naegleria fowleri* excretory-secretory proteins and their biochemical properties. Parasitol. Res. 113, 2765-2776.

Leippe, M., and Herbst, R. (2004). Ancient weapons for attack and defense: the pore-forming polypeptides of pathogenic enteric and free-living amoeboid protozoa. J. Eukaryot. Microbiol. 51, 516-521.

Levy, Y.Y., Lai, E.Y., Remillard, S.P., Heintzelman, M.B., and Fulton, C. (1996). Centrin is a conserved protein that forms diverse associations with centrioles and MTOCs in *Naegleria* and other organisms. Cell Motil. Cytoskeleton 33, 298-323.

Levy, Y.Y., Lai, E.Y., Remillard, S.P., and Fulton, C. (1998). Centrin is synthesized and assembled into basal bodies during *Naegleria* differentiation. Cell Motil. Cytoskeleton 40, 249-260.

Linam, W.M., Ahmed, M., Cope, J.R., Chu, C., Visvesvara, G.S., da Silva, A.J., Qvarnstrom, Y., and Green, J. (2015). Successful treatment of an adolescent with *Naegleria fowleri* primary amoebic meningoencephalitis. Pediatrics 135, e744-748.

Lopez, C., Budge, P., Chen, J., Bilyeu, S., Mirza, A., Custodio, H., Irazuzta, J., Visvesvara, G., and Sullivan, K.J. (2012). Primary amoebic meningoencephalitis: a case report and literature review. Pediatr. Emerg. Care 28, 272-276.

Lowrey, D.M., and McLaughlin, J. (1984). A multicomponent haemolytic system in the pathogenic amoeba *Naegleria fowleri*. Infect. Immun. 45, 731-736.

Lowrey, D.M., and McLaughlin, J. (1985). Subcellular distribution of hydrolases in *Naegleria fowleri*. J. Protozool. 32, 616-621.

Lowrey, D.M., and McLaughlin, J. (1985). Activation of a heat-stable cytolytic protein associated with the surface membrane of *Naegleria fowleri*. Infect. Immun. 50, 478-482.

Lyu, C.J., Shin, C.O., and Im, K. (1993). Interleukin-2 production and alteration of T cell subsets in mice infected with *Naegleria fowleri*. Kor. J. Parasitol. 31, 249-257.

Maclean, R.C., Richardson, D.J., LePardo, R., and Marciano-Cabral, F. (2004). The identification of *Naegleria fowleri* from water and soil samples by nested PCR. Parasitol. Res. 93, 211-217.

Madarová, L., Trnková, K., Feiková, S., Klement, C., and Obernauerová, M. (2010). A real-time PCR diagnostic method for detection of *Naegleria fowleri*. Exp. Parasitol. 126, 37-41.

Mahittikorn, A., Mori, H., Popruk, S., Roobthaisong, A., Sutthikornchai, C., Koompapong, K., Siri, S., Sukthana, Y., and Nacapunchai, D. (2015). Development of a rapid, simple method for detecting *Naegleria fowleri* visually in water samples by Loop-Mediated Isothermal Amplification (LAMP). PLoS One 10, e0120997.

Maitra, S.C., Krishna-Prasad, B.N., Das, S.R., and Agarwala, S.C. (1974). Study of *Naegleria aerobia* by electron microscopy. Trans. R. Soc. Trop. Med. Hyg. 68, 56-60.

Mar, J., Lee, J.H., Shea, D., and Walsh, C.J. (1986). New poly(A)+RNAs appear coordinately during the differentiation of *Naegleria gruberi* amoebae into flagellates. J. Cell Biol. 102, 353-361.

Mar, J., and Walsh, C.J. (2007). Temperature-shock induction of multiple flagella induces additional synthesis of flagellum specific mRNAs and tubulin. Exp. Cell Res. 314, 896-902.

Marciano-Cabral, F.M., Patterson, M., John, D.T., and Bradley, S.G. (1982). Cytopathogenicity of *Naegleria fowleri* and *Naegleria gruberi* for established mammalian cell cultures. J. Parasitol. 68, 1110-1116.

Marciano-Cabral, F., and John, D.T. (1983). Cytopathogenicity of *Naegleria fowleri* for rat neuroblastoma cell cultures: scanning electron microscopy study. Infect. Immun. 40, 1214-1217.

Marciano-Cabral, F., Cline, M.L., and Bradley, S.G. (1987). Specificity of antibodies from human sera for *Naegleria* species. J. Clin. Microbiol. 25, 692-697.

Marciano-Cabral, F., and Cline, M. (1987). Chemotaxis by *Naegleria fowleri* for bacteria. J. Protozool. 34, 127-131.

Marciano-Cabral, F., Stanitski, S., Radhakrishna, V., and Bradley, S.G. (1987). Characterization of a neutral aminoacyl-peptide hydrolase from *Naegleria fowleri*. J. Protozool. 34, 146-149.

Marciano-Cabral, F., Zoghby, K.L., and Bradley, S.G. (1990). Cytopathic action of *Naegleria fowleri* amoebae on rat neuroblastoma target cells. J. Protozool. 37, 138-144.

Marciano-Cabral, F., and Toney, D.M. (1994). Modulation of biological functions of *Naegleria fowleri* amoebae by growth medium. J. Eukaryot. Microbiol. 41, 38-46.

Marciano-Cabral, F., MacLean, R., Mensah, A., and LaPat-Polasko, L. (2003). Identification of *Naegleria fowleri* in domestic water sources by nested PCR. Appl. Environ. Microbiol. 69, 5864-5869.

Marciano-Cabral, F., and Cabral, G.A. (2007). The immune response to *Naegleria fowleri* amoebae and pathogenesis of infection. FEMS Immunol. Med. Microbiol. 51, 243-259.

Markowitz, S.M., Martinez, A.J., Duma, R.J., Shiel, F.O.M. (1974). Myocarditis associated with primary amoebic (*Naegleria*) meningoencephalitis. Am. J. Clin. Pathol. 62, 619-628.

Marschner, N., Kotting, J., Eibl, H., and Unger, C. (1992). Distribution of hexadecylphosphocholine and octadecyl-methyl-glycero-3-phosphocholine in rat tissues during steady-state treatment. Cancer Chemother. Pharmacol. 31, 18-22.

Martin, L.H., Corff, S.C., and Yuyama, S. (1982). DNA synthesis during growth and synchronous differentiation of *Naegleria*. Cytobios. 33, 29-38.

Martínez, A.J. (1977). Free-living amoebic meningo-encephalitides: comparative study. Neurol. Neurocir. Psiquiatr. 18, (2-3 Suppl), 391-401.

Martinez, A.J. (1985). Free-Living Amoebas: Natural History, Prevention, Diagnosis, Pathology, and Treatment of Disease. Boca Ralton, FL, USA, CRC Press.

Maruyama, S., and Nozaki, H. (2007). Sequence and intranuclear location of the extrachromosomal rDNA plasmid of the amoebo-flagellate *Naegleria gruberi*. J. Eukaryot. Microbiol. 54, 333-337.

Mat Amin, N. (2004). Proteinases in *Naegleria fowleri* (strain NF3), a pathogenic amoeba: a preliminary study. Trop. Biomed. 21, 57-60.

May, R.G., and John, D.T. (1982). Intravenous infection of mice with *Naegleria fowleri*. Folia Parasitol. (Praha). 29, 201-210.

Mazzola, M., de Bruijn, I., Cohen, M.F., and Raaijmakers, J.M. (2009). Protozoan-induced regulation of cyclic lipopeptide biosynthesis is an effective predation defense mechanism for *Pseudomonas fluorescens*. Appl. Environ. Microbiol. 75, 6804-6811.

McLaughlin, G.L., Vodkin, M.H., and Huizinga, H.W. (1991). Amplification of repetitive DNA for the specific detection of *Naegleria fowleri*. J. Clin. Microbiol. 29, 227-230.

Merchant, M., Thibodeaux, D., Loubser, K., Elsey, R.M. (2004). Amoebicidal effects of serum from the American alligator (*Alligator mississippiensis*). J. Parasitol. 90, 1480-1483.

Mertens, E., De Jonckheere, J., and Van Schaftingen, E. (1993). Pyrophosphate-dependent phosphofructokinase from the amoeba *Naegleria fowleri*, an AMP-sensitive enzyme. Biochem. J. 292, 797-803.

Michel, R., and Just, H.M. (1984). *Acanthamoebae*, *Naegleria* and other free-living amoebae in cooling and rinsing water of dental treatment units. Zentralbl. Bakteriol. Mikrobiol. Hyg. B. 179, 56-72.

Michel, R., and Borneff, M. (1989). The significance of amoebae and other protozoa in water conduit systems in dental units. Zentralbl. Bakteriol. Mikrobiol. Hyg. B. 187, 312-323.

Michel, R., and Menn, T. (1991). *Acanthamoeba*, *Naegleria* and invertebrates in wet areas of physiotherapy equipment in hospitals. Zentralbl. Hyg. Umweltmed. 191, 423-437.

Michel, R., Hauroder, B., Muller, K.D., Zoller, L. (1999). An environmental *Naegleria* strain, unable to form cysts turned out to harbor two different species of endosymbionts. Endocytobios. Cell Res. 115-118.

Michel, R., Müller, K.D., Hauröder, B., Zöller, L. (2000). A coccoid bacterial parasite of *Naegleria* sp. (Schizopyrenida: Vahlkampfiidae) inhibits cyst formation of its host but not transformation to the flagellate stage. Acta Protozool. 39, 199-207.

Michelson, M.K., Henderson, W.R.Jr., Chi, E.Y., Fritsche, T.R., and Klebanoff, S.J. (1990). Ultrastructural studies on the effect of tumor necrosis factor on the interaction of neutrophils and *Naegleria fowleri*. Am. J. Trop. Med. Hyg. 42, 225-233.

Morahan, P.S., Bradley, S.G., Munson, A.E., Duke, S., Fromtling, R.A., and Marciano-Cabral, F. (1984). Immunotoxic effects of diethylstilbestrol on host resistance: comparison with cyclophosphamide. J. Leukoc. Biol. 35, 329-341.

Moss, D.M., Brandt, F.H., Mathews, H.M., and Visvesvara, G.S. (1988). High-resolution polyacrylamide gradient gel electrophoresis (PGGE) of isoenzymes from five *Naegleria* species. J. Protozool. 35, 26-31.

Moura, H., Izquierdo, F., Woolfitt, A.R., Wagner, G., Pinto, T., Del Aguila, C., and Barr, J.R. (2014). Detection of biomarkers of pathogenic *Naegleria fowleri* through mass spectrometry and proteomics. J. Eukaryot. Microbiol. 62, 12-20.

Moussa, M., Tissot, O., Guerlotté, J., De Jonckheere, J.F., and Talarmin, A. (2014). Soil is the origin for the presence of *Naegleria fowleri* in the thermal recreational waters. Parasitol. Res. 114, 311-315.

Muldrow, L.L., Tyndall, R.L., and Fliermans, C.B. (1982). Application of flow cytometry to studies of pathogenic free-living amoebae. Appl. Environ. Microbiol. 44, 1258-1269.

Mull, B.J., Narayanan, J., and Hill, V.R. (2013). Improved method for the detection and quantification of *Naegleria fowleri* in water and sediment using immunomagnetic separation and real-time PCR. J. Parasitol. Res. e608367.

Nerad, T.A., Visvesvara, G., and Daggett, P.M. (1983). Chemically defined media for the cultivation of *Naegleria*: pathogenic and high temperature tolerant species. J. Protozool. 30, 383-387.

Newsome, A.L., and Wilhelm, W.E. (1981). Effect of exogenous iron on the viability of pathogenic *Naegleria fowleri* in serum. Experientia 37, 1160-1162.

Newsome, A.L., and Wilhelm, W.E. (1983). Inhibition of *Naegleria fowleri* by microbial iron-chelating agents: ecological implications. Appl. Environ. Microbiol. 45, 665-668.

Newsome, A.L., and Arnold, R.R. (1985). Equivalent mortality in normal and athymic mice infected with *Naegleria fowleri*. J. Parasitol. 71, 678-679.

Newsome, A.L., Baker, R.L., Miller, R.D., and Arnold, R.R. (1985). Interactions between *Naegleria fowleri* and *Legionella pneumophila*. Infect. Immun. 50, 449-452.

O'Dell, W.D., and Stevens, A.R. (1973). Quantitative growth of *Naegleria* in axenic culture. Appl. Microbiol. 25, 621-627.

O'Dell, W.D., and Brent, M.M. (1974). Nutritional study of three strains of *Naegleria gruberi*. J. Protozool. 21, 129-133.

Oh, Y.H., Jeong, S.R., Kim, J.H., Song, K.J., Kim, K., Park, S., Sohn, S., and Shin, H.J. (2005). Cytopathic changes and pro-inflammatory cytokines induced by *Naegleria fowleri* trophozoites in rat microglial cells and protective effects of an anti-Nfa1 antibody. Parasite Immunol. 27, 453-459.

Olomu, N., Martinez, A.J., Lamarco, K.L., Nerad, T.A., Saha, A.K., Das, S., and Glew, R.H. (1986). Demonstration of various acid hydrolases and preliminary characterization of acid phosphatase in *Naegleria fowleri*. J. Protozool. 33, 317-321.

Ondarza, R.N., Iturbe, A., Hernández, E., and Hurtado, G. (2003). Low-molecular-mass thiol compounds from a free-living highly pathogenic amoeba, *Naegleria fowleri*. Biotechnol. Appl. Biochem. 37, 195-204.

Ondarza, R.N., Hurtado, G., Tamayo, E., Iturbe, A., and Hernández, E. (2006). *Naegleria fowleri*: a free-living highly pathogenic amoeba contains trypanothione/trypanothione reductase and glutathione/glutathione reductase systems. Exp. Parasitol. 114, 141-146.

Ondarza, R.N., Iturbe, A., and Hernandez, E. (2006). *In vitro* antiproliferative effects of neuroleptics, antimycotics and antibiotics on the human pathogens *Acanthamoeba polyphaga* and *Naegleria fowleri*. Arch. Med. Res. 37, 723-729.

Ondarza, R.N. (2007). Drug targets from human pathogenic amoebas: *Entamoeba histolytica*, *Acanthamoeba polyphaga* and *Naegleria fowleri*. Infect. Disord. Drug Targets 7, 266-280.

Ondarza, R.N., Iturbe, A., and Hernández, E. (2006). *In vitro* antiproliferative effects of neuroleptics, antimycotics and antibiotics on the human pathogens *Acanthamoeba polyphaga* and *Naegleria fowleri*. Arch. Med. Res. 37, 723-729.

Ondarza, R.N., Iturbe, A., and Hernández, E. (2007). The effects by neuroleptics, antimycotics and antibiotics on disulfide reducing enzymes from the human pathogens *Acanthamoeba polyphaga* and *Naegleria fowleri*. Exp. Parasitol. 115, 41-47.

Ondarza, R.N. (2008). Drug effects on drug targets: inhibition of enzymes by neuroleptics, antimycotics, antibiotics and other drugs on human pathogenic amoebas and their anti-proliferative effects. Recent. Pat. Antiinfect. Drug. Discov. 2, 206-216.

Opperdoes, F.R., De Jonckheere, J.F., and Tielens, A.G. (2011). *Naegleria gruberi* metabolism. Int. J. Parasitol. 41, 915-924.
Osuna, A., Rodriguez-Santiago, J.I., Ruiz-Perez, L.M., Gamarro, F., Castanys, S., Giovannangeli, G., Galy, A.M., Galy, J.P., Soyfer, J.C., and Barbe, J. (1987). Antiamoebic activity of new acridinic derivatives against *Naegleria* and *Acanthamoeba* species *in vitro*. Chemother. 33, 18-21.
Park, K.M., Ryu, J.S., and Im, K.I. (1987). Blastogenic responses of splenic lymphocytes to *Naegleria fowleri* lysates and T-cell mitogen in mice with primary amoebic meningoencephalitis. Kisaengchunghak Chapchi. 25, 1-6.
Patterson, M., Woodworth, T.W., Marciano-Cabral, F., and Bradley, S.G. (1981). Ultrastructure of *Naegleria fowleri* enflagellation. J. Bacteriol. 147, 217-226.
Pearson, J.L., and Willmer, E.N., 1963. Some observations on the actions of steroids on the metaplasia of the amoeba, *Naegleria gruberi*. J. Exp. Biol. 40, 493-515.
Pélandakis, M., De Jonckheere, J.F., and Pernin, P. (1998). Genetic variation in the free-living amoeba *Naegleria fowleri*. Appl. Environ. Microbiol. 64, 2977-2981.
Pélandakis, M., Serre, S., and Pernin, P. (2000). Analysis of the 5.8S rRNA gene and the internal transcribed spacers in *Naegleria* spp. and in *N. fowleri*. J. Eukaryot. Microbiol. 47, 116-121.
Pélandakis, M., and Pernin, P. (2002). Use of multiplex PCR and PCR restriction enzyme analysis for detection and exploration of the variability in the free-living amoeba *Naegleria* in the environment. Appl. Environ. Microbiol. 68, 2061-2065.
Pernin, P., Bouikhsain, I., De Jonckheere, J.F., and Petavy, A.F. (1983). Comparative protein patterns of three thermophilic nonpathogenic *Naegleria* isolates and two *Naegleria fowleri* strains by isoelectric focusing. Int. J. Parasitol. 13, 113-118.
Pernin, P. (1984). Isoenzyme patterns of pathogenic and nonpathogenic thermophilic *Naegleria* strains by isoelectric focusing. Int. J. Parasitol. 14, 459-465.
Pernin, P., Cariou, M.L., and Jacquier, A. (1985). Biochemical identification and phylogenetic relationships in free-living

amoebas of the genus *Naegleria*. J. Protozool. 32, 592-603.

Pernin, P., and Grelaud, G. (1989). Application of isoenzymatic typing to the identification of nonaxenic strains of *Naegleria* (Protozoa, Rhizopoda). Parasitol. Res. 75, 595-598.

Pernin, P., Pélandakis, M., Rouby, Y., Faure, A., and Siclet, F. (1998). Comparative recoveries of *Naegleria fowleri* amoebae from seeded river water by filtration and centrifugation. Appl. Environ. Microbiol. 64, 955-959.

Perrine, D., Barbier, D., and Georges, P. (1980). Comparative *in vitro* activity of chlorine and bromine on the cysts of free-living amoebae. C R Seances Soc. Biol. Fil. 174, 297-303.

Phillips, B.P. (1974). *Naegleria*: another pathogenic amoeba studies in germfree guinea pigs. Am. J. Trop. Med. Hyg. 23, 850-855.

Phu, N.H., Hoang Mai, N.T., Nghia, H.D., Chau, T.T., Loc, P.P., Thai le, H., Phuong, T.M., Thai, C.Q., Man, D.N., Van Vinh Chau, N., Nga, T.V., Campbell, J., Baker, S., and Whitehorn, J. (2013). Fatal consequences of freshwater pearl diving. Lancet 381(9861), 176.

Pougnard, C., Catala, P., Drocourt, J.L., Legastelois, S., Pernin, P., Pringuez, E., and Lebaron, P. (2002). Rapid detection and enumeration of *Naegleria fowleri* in surface waters by solid-phase cytometry. Appl. Environ. Microbiol. 68, 3102-3107.

Powell, E.L., Newsome, A.L., Allen, S.D., and Knudson, G.B. (1994). Identification of antigens of pathogenic free-living amoebae by protein immunoblotting with rabbit immune and human sera. Clin. Diagn. Lab. Immunol. 1, 493-499.

Preston, T.M., and O'Dell, D.S. (1974). Studies on the amoeboid and flagellate stages of *Naegleria*. Ann. Soc. Belg. Med. Trop. 54, 279-286.

Preston, T.M., and King, C.A. (1978). An experimental study of the interaction between the soil amoeba *Naegleria gruberi* and a glass substrate during amoeboid locomotion. J. Cell Sci. 34, 145-158.

Preston, T.M., Cooper, L.G., and King, C.A. (1990). Amoeboid locomotion of *Naegleria gruberi*: the effects of

cytochalasin B on cell-substratum interactions and motile behavior. J. Protozool. 37, 6S-11S.
Preston, T.M., and King, C.A. (2003). Locomotion and phenotypic transformation of the amoeboflagellate *Naegleria gruberi* at the water-air interface. J. Eukaryot. Microbiol. 50, 245-251.
Prevention CfDCa. (2013). Investigational drug available directly from CDC for the treatment of infections with free-living amoebae. Morb. Mort. Week. Rep. 62(33), 666.
Pringle, H.L., Bradley, S.G., and Harris, L.S. (1979). Susceptibility of *Naegleria fowleri* to delta 9-tetrahydrocannabinol. Antimicrob. Agents Chemother. 16, 674-679.
Puzon, G.J., Lancaster, J.A., Wylie, J.T., and Plumb, I.J. (2009). Rapid detection of *Naegleria fowleri* in water distribution pipeline biofilms and drinking water samples. Environ. Sci. Technol. 43, 6691-6696.
Qvarnstrom, Y., Visvesvara, G.S., Sriram, R., and da Silva, A.J. (2006). Multiplex real-time PCR assay for simultaneous detection of *Acanthamoeba* spp., *Balamuthia mandrillaris*, and *Naegleria fowleri*. J. Clin. Microbiol. 44, 3589-3595.
Raederstorff, D., and Rohmer, M. (1987). Sterol biosynthesis via cycloartenol and other biochemical features related to photosynthetic phyla in the amoeba *Naegleria lovaniensis* and *Naegleria gruberi*. Eur. J. Biochem. 164, 427-434.
Ramirez, E., Robles, E., and Martinez, B. (2010). Free-living amoebae isolated from water-hyacinth root (*Eichhornia crassipes*). Exp. Parasitol. 126, 42-44.
Ramirez, E., Robles, E., Martinez, B., Ayala, R., Sainz, G., Martinez, M.E., and Gonzalez, M.E. (2014). Distribution of free-living amoebae in a treatment system of textile industrial wastewater. Exp. Parasitol. 145, Suppl:S34-38.
Rasmussen, J.P., Cursaro, M., Froscio, S.M., and Saint, C.P. (2008). An examination of the antibiotic effects of cylindrospermopsin on common gram-positive and gram-negative bacteria and the protozoan *Naegleria lovaniensis*. Environ. Toxicol. 23, 36-43.

Ravine, T.J., Polski, J.M., and Jenkins, J. (2010). Recognition of *Naegleriae* amoeba surface protein epitopes by anti-human CD45 antibodies. Cytometry A. 77, 305-309.

Reilly, M.F., Bradley, M.K., and Bradley, S.G. (1982). Agglutination of *Naegleria fowleri* by human serum. Proc. Soc. Exp. Biol. Med. 170, 209-212.

Reilly, M.F., Marciano-Cabral, F., Bradley, D.W., and Bradley, S.G. (1983). Agglutination of *Naegleria fowleri* and *Naegleria gruberi* by antibodies in human serum. J. Clin. Microbiol. 17, 576-581.

Reilly, M.F., White, K.L.Jr., and Bradley, S.G. (1983). Host resistance of mice to *Naegleria fowleri* infections. Infect. Immun. 42, 645-652.

Remillard, S.P., Lai, E.Y., Levy, Y.Y., and Fulton, C. (1995). A calcineurin-B-encoding gene expressed during differentiation of the amoeboflagellate *Naegleria gruberi* contains two introns. Gene 154, 39-45.

Réveiller, F.L., Marciano-Cabral, F., Pernin, P., Cabanes, P.A., and Legastelois, S. (2000). Species specificity of a monoclonal antibody produced to *Naegleria fowleri* and partial characterization of its antigenic determinant. Parasitol. Res. 86, 634-641.

Réveiller, F.L., Suh, S.J., Sullivan, K., Cabanes, P.A., and Marciano-Cabral, F. (2001). Isolation of a unique membrane protein from *Naegleria fowleri*. J. Eukaryot. Microbiol. 48, 676-682.

Réveiller, F.L., Cabanes, P.A., and Marciano-Cabral, F. (2002). Development of a nested PCR assay to detect the pathogenic free-living amoeba *Naegleria fowleri*. Parasitol. Res. 88, 443-450.

Reveiller, F.L., Varenne, M.P., Pougnard, C., Cabanes, P.A., Pringuez, E., Pourima, B., Legastelois, S., and Pernin, P. (2003). An enzyme-linked immunosorbent assay (ELISA) for the identification of *Naegleria fowleri* in environmental water samples. J. Eukaryot. Microbiol. 50, 109-113.

Rice, C.A., Colon, B.L., Alp, M., Göker, H., Boykin, D.W., and Kyle, D.E. (2015). Bis-Benzimidazole Hits against *Naegleria fowleri* Discovered with New High-Throughput Screens. Antimicrob. Agents Chemother. 59, 2037-2044.

Rivera, F., Galván, M., Robles, E., Leal, P., González, L., and Lacy, A.M. (1981). Bottled mineral waters polluted by protozoa in Mexico. J. Protozool. 28, 54-56.

Rivera, F., Ramírez, P., Vilaclara, G., Robles, E., and Medina, F. (1983). A survey of pathogenic and free-living amoebae inhabiting swimming pool water in Mexico City. Environ. Res. 32, 205-211.

Rivera, F., Medina, F., Ramírez, P., Alcocer, J., Vilaclara, G., and Robles, E. (1984). Pathogenic and free-living protozoa cultured from the nasopharyngeal and oral regions of dental patients. Environ. Res. 33, 428-440.

Rivera, F., Roy-Ocotla, G., Rosas, I., Ramirez, E., Bonilla, P., and Lares, F. (1987). Amoebae isolated from the atmosphere of Mexico City and environs. Environ. Res. 42, 149-154.

Rivera, F., Ramírez, E., Bonilla, P., Calderón, A., Gallegos, E., Rodríguez, S., Ortiz, R., Zaldívar, B., Ramírez, P., and Durán, A. (1993). Pathogenic and free-living amoebae isolated from swimming pools and physiotherapy tubs in Mexico. Environ. Res. 62, 43-52.

Rivera-Aguilar, V., Hernández-Martínez, D., Rojas-Hernández, S., Oliver-Aguillón, G., Tsutsumi, V., Herrera-González, N., and Campos-Rodríguez, R. (2000). Immunoblot analysis of IgA antibodies to *Naegleria fowleri* in human saliva and serum. Parasitol. Res. 86, 775-780.

Rivera, V., Hernández, D., Rojas, S., Oliver, G., Serrano, J., Shibayama, M., Tsutsumi, V., and Campos, R. (2001). IgA and IgM anti-*Naegleria fowleri* antibodies in human serum and saliva. Can. J. Microbiol. 47, 464-466.

Robinson, B.S., Monis, P.T., and Dobson, P.J. (2006). Rapid, sensitive, and discriminating identification of *Naegleria* spp. by real-time PCR and melting-curve analysis. Appl. Environ. Microbiol. 72, 5857-5863.

Rodriguez-Monroy, M.A., and Moreno-Fierros, L. (2010). Striking activation of NALT and nasal passages lymphocytes induced by intranasal immunization with Cry1Ac protoxin. Scand. J. Immunol. 71, 159-168.

Rojas-Hernández, S., Rodríguez-Monroy, M.A., López-Revilla, R., Reséndiz-Albor, A.A., and Moreno-Fierros, L. (2004). Intranasal coadministration of the Cry1Ac protoxin with amoebal lysates increases protection against *Naegleria fowleri* meningoencephalitis. Infect. Immun. 72, 4368-4375.

Rojas-Hernández, S., Jarillo-Luna, A., Rodríguez-Monroy, M., Moreno-Fierros, L., and Campos-Rodríguez, R. (2004). Immunohistochemical characterization of the initial stages of *Naegleria fowleri* meningoencephalitis in mice. Parasitol. Res. 94, 31-36.

Rojas-Hernández, S., Rodríguez-Monroy, M.A., Moreno-Fierros, L., Jarillo-Luna, A., Carrasco-Yepez, M., Miliar-García, A., and Campos-Rodríguez, R. (2007). Nitric oxide production and nitric oxide synthase immunoreactivity in *Naegleria fowleri*. Parasitol. Res. 101, 269-274.

Rowan-Kelly, B., Ferrante, A., and Thong, Y.H. (1980). Activation of complement by *Naegleria*. Trans. R. Soc. Trop. Med. Hyg. 74, 333-336.

Rowbury, J., Armitage, J.P., King, C. (1983). Movement, taxes and cellular interactions in the response of microorganisms to their natural environment. p. 299-35t. In J. H. Slater. R. Whittenbury, and J. W. T. Wimpenny (ed.), Microbes in their natural environments. vol. 34. Cambridge University Press, Cambridge.

Rüdinger, M., Fritz-Laylin, L., Polsakiewicz, M., and Knoop, V. (2011). Plant-type mitochondrial RNA editing in the protist *Naegleria gruberi*. RNA 17, 2058-2062.

Ryu, J.S., Soh, C.T., and Im, K.I. (1984). Ultrastructural observation of *Naegleria fowleri* trophozoite in mouse brain and axenic culture. Kisaengchunghak Chapchi. 22, 259-266.

Ryu, J.S., and Im, K.I. (1992). The production and characterization of anti-*Naegleria fowleri* monoclonal antibodies. Kisaengchunghak Chapchi. 30, 33-41.

Saanen, C.V. (1980). Comparative scanning electron microscopic study of *Naegleria* and *Acanthamoeba* (*Hartmannella*) cysts. Ann. Parasitol. Hum. Comp. 55, 645-657.

Sateriale, A., Bessoff, K., Sarkar, I.N., and Huston, C.D. (2014). Drug repurposing: mining protozoan proteomes for targets of known bioactive compounds. J. Am. Med. Inform. Assoc. 21, 238-244.

Scaglia, M., Strosselli, M., Grazioli, V., Gatti, S., Bernuzzi, A.M., and De Jonckheere, J.F. (1983). Isolation and identification of pathogenic *Naegleria australiensis* (Amoebida, Vahlkampfiidae) from a spa in northern Italy. Appl. Environ. Microbiol. 46, 1282-1285.

Scaglia, M., Gatti, S., Brustia, R., Strosselli, M., Bernuzzi, A.M., and Cevini, C. (1987). Pathogenic and non-pathogenic *Naegleria* and *Acanthamoeba* spp.: a new autochthonous isolate from an Italian thermal area. Microbiologica 10, 171-182.

Scaglia, M., Gatti, S., Bernuzzi, A.M., Cevini, C., Chichino, G., and Rondanelli, E.G. (1988). An *in vitro* comparative study on the effect of amphotericin B, econazole, and 5-fluorocytosine on *Naegleria fowleri*, *Naegleria australiensis*, and *Naegleria australiensis* s.sp. *italica*. Microbiologica 11, 279-288.

Scaglia, M., Gatti, S., Cevini, C., Bernuzzi, A.M., and Martinez, A.J. (1989). *Naegleria australiensis* ssp. *italica*: experimental study in mice. Exp. Parasitol. 69, 294-299.

Scaglia, M., Gatti, S., Brustia, R., Chichino, G., and Rondanelli, E.G. (1991). Phagocytosis of human erythrocytes by *Naegleria* is not related to species pathogenicity. A phase-contrast cinemicrographic study. Microbiologica 14, 45-53.

Schild, M., Gianinazzi, C., Gottstein, B., and Müller, N. (2007). PCR-based diagnosis of *Naegleria* sp. infection in formalin-fixed and paraffin-embedded brain sections. J. Clin. Microbiol. 45, 564-567.

Schuster, F. (1963). An electron microscope study of the amoeba-flagellate, *Naegleria gruberi* (Schardinger). I. The amoeboid and flagellate stages. J. Protozool. 10, 297-313.

Schuster, F. (1963). An electron microscope study of the amoeba-flagellate, *Naegleria gruberi* (Schardinger). II. The cyst stage. J. Protozool. 10, 313-320.

Schuster, F.L., and Svihla, G. (1968). Ribonucleoprotein-containing vesicles in cysts of *Naegleria gruberi*. J. Protozool. 15, 752-758.

Schuster, F.L. (1969). Intranuclear virus-like bodies in the amoeboflagellate *Naegleria gruberi*. J. Protozool. 16, 724-727.

Schuster, F.L. (1974). Ultrastructure of mitosis in the amoeboflagellate *Naegleria gruberi*. Tissue Cell. 7, 1-11.
Schuster, F.L., and Dunnebacke, T.H. (1974). Growth at 37°C of the EGs strain of the amoeboflagellate *Naegleria gruberi* containing viruslike particles. I. Nuclear changes. J. Invertebr. Pathol. 23, 172-181.
Schuster, F.L., and Dunnebacke, T.H. (1974). Growth at 37°C of the EGs strain of the amoeboflagellate *Naegleria gruberi* containing viruslike particles. II. Cytoplasmic changes. J. Invertebr. Pathol. 23, 182-189.
Schuster, F.L., and Rechthand, E. (1975). *In vitro* effects of amphotericin B on growth and ultrastructure of the amoeboflagellates *Naegleria gruberi* and *Naegleria fowleri*. Antimicrob. Agents Chemother. 8, 591-605.
Schuster, F.L., and Clemente, J.S. (1977). 5-Bromodeoxyuridine-induced formation of virus-like particles in *Naegleria gruberi* EGs. J. Cell Sci. 26, 359-371.
Schuster, F.L., and Dunnebacke, T.H. (1977). Ultrastructural observations of experimental *Naegleria* meningo-encephalitis in mice: intranuclear inclusions in amoebae and host cells. J. Protozool. 24, 489-497.
Schuster, F.L., and Twomey, R. (1983). Calcium regulation of flagellation in *Naegleria gruberi*. J. Cell Sci. 63, 311-326.
Schuster, F.L., and Mandel, N. (1984). Phenothiazine compounds inhibit *in vitro* growth of pathogenic free-living amoebae. Antimicrob. Agents Chemother. 25, 109-112.
Schuster, F.L., and Visvesvara, G.S. (1998). Efficacy of novel antimicrobials against clinical isolates of opportunistic amoebas. J. Eukaryot. Microbiol. 45, 612-618.
Schuster, F.L., and Visvesvara, G.S. (2001). Activity of selected antimicrobials against clinical isolates of pathogenic free-living amoebas. Paris, France: John Libbey Eurotext.
Schuster, F.L. (2002). Cultivation of pathogenic and opportunistic free-living amoebas. Clin. Microbiol. Rev. 15, 342-354.
Schuster, F.L., and Visvesvara, G.S. (2004). Free-living amoebae as opportunistic and non-opportunistic pathogens of humans and animals. Int. J. Parasitol. 34, 1001-1027.

Schuster, F.L., Guglielmo, B.J., and Visvesvara, G.S. (2006). *In vitro* activity of miltefosine and voriconazole on clinical isolates of free-living amoebas: *Balamuthia mandrillaris*, *Acanthamoeba* spp., and *Naegleria fowleri*. J. Eukaryot. Microbiol. 53, 121-126.

Seidel, J.S., Harmatz, P., Visvesvara, G.S., Cohen, A., Edwards, J., and Turner, J. (1982). Successful treatment of primary amoebic meningoencephalitis. N. Engl. J. Med. 306, 346-348.

Serrano-Luna, J., Cervantes-Sandoval, I., Tsutsumi, V., and Shibayama, M. (2007). A biochemical comparison of proteases from pathogenic *Naegleria fowleri* and non-pathogenic *Naegleria gruberi*. J. Eukaryot. Microbiol. 54, 411-417.

Sesma, M.J., and Ramos, L.Z. (1989). Isolation of free-living amoebas from the intestinal contents of reptiles. J. Parasitol. 75, 322-324.

Shea, D.K., and Walsh, C.J. (1987). mRNAs for alpha- and beta-tubulin and flagellar calmodulin are among those coordinately regulated when *Naegleria gruberi* amoebae differentiate into flagellates. J. Cell Biol. 105, 1303-1309.

Sheehan, K.B., Fagg, J.A., Ferris, M.J., Henson, J.M. (2003). PCR detection and analysis of the free-living amoeba *Naegleria* in hot springs in Yellowstone and Grand Teton National Parks. Appl. Environ. Microbiol. 69, 5914-5918.

Shibayama, M., Serrano-Luna Jde, J., Rojas-Hernández, S., Campos-Rodríguez, R., and Tsutsumi, V. (2003). Interaction of secretory immunoglobulin A antibodies with *Naegleria fowleri* trophozoites and collagen type I. Can. J. Microbiol. 49, 164-170.

Shibayama, M., Martínez-Castillo, M., Silva-Olivares, A., Galindo-Gómez, S., Navarro-García, F., Escobar-Herrera, J., Sabanero, M., Tsutsumi, V., and Serrano-Luna, J. (2013). Disruption of MDCK cell tight junctions by the free-living amoeba *Naegleria fowleri*. Microbiol. 159, 392-401.

Shin, H.J., Im, K.I., and Choe, R.S. (1985). Effect of splenectomy on development of primary amoebic meningoencephalitis. Kisaengchunghak Chapchi. 23, 156-164.

Shin, H.J., Cho, M.S., Jung, S.U., Kim, H.I., Park, S., Kim, H.J., and Im, K.I. (2001). Molecular cloning and characterization of a gene encoding a 13.1 kDa antigenic protein of *Naegleria fowleri*. J. Eukaryot. Microbiol. 48, 713-717.

Siddiqui, R., and Khan, N. A. (2012). Is ritual cleansing a missing link between fatal infection and brain eating amoeba? Clin. Infect. Dis. 54, 1817-1818.

Siddiqui, R., Khan, N. A. (2014). Primary amoebic meningoencephalitis caused by *Naegleria fowleri*: an old enemy presenting new challenges. PLoS NTD 8, e3017.

Simpson, C.F., Willaert, E., Neal, F.C., Stevens, A.R., and Young, M.D. (1982). Experimental *Naegleria fowleri* meningoencephalitis in sheep: light and electron microscopic studies. Am. J. Vet. Res. 43, 154-157.

Singh, S.N., Patwari, A.K., Dutta, R., Taneja, N., and Anand, V.K. (1998). *Naegleria* meningitis. Indian Pediatr. 35, 1012-1015.

Siripanth, C., Punpoowong, B., and Riganti, M. (2005). Early detection and identification of amphizoic amoebae from nasal exudates of a symptomatic case. J. Med. Assoc. Thai. 88, 545-549.

Smego, R.A.Jr., and Durack, D.T. (1984). An experimental model for *Naegleria fowleri*-induced primary amoebic meningoencephalitis in rabbits. J. Parasitol. 70, 78-81.

Soh, E.Y., Shin, H.J., and Im, K. (1992). The protective effects of monoclonal antibodies in mice from *Naegleria fowleri* infection. Kisaengchunghak Chapchi. 30, 113-123.

Sohn, H.J., Kim, J.H., Shin, M.H., Song, K.J., and Shin, H.J. (2010). The *Nf-actin* gene is an important factor for food-cup formation and cytotoxicity of pathogenic *Naegleria fowleri*. Parasitol. Res. 106, 917-924.

Soll, D.R., and Fulton, C. (1974). The constancy of RNA-polymerase activities during a transcriptionally dependent cell differentiation in *Naegleria*. Dev. Biol. 36, 236-244.

Solomon, L.R., and Rubenstein, P.A. (1987). Studies on the role of actin's N tau-methylhistidine using oligodeoxy-nucleotide-directed site-specific mutagenesis. J. Biol. Chem. 262, 11382-11388.

Soltow, S.M., and Brenner, G.M. (2007). Synergistic activities of azithromycin and amphotericin B against *Naegleria fowleri in vitro* and in a mouse model of primary amoebic meningoencephalitis. Antimicrob. Agents Chemother. 51, 23-27.

Song, K.J., Jeong, S.R., Park, S., Kim, K., Kwon, M.H., Im, K.I., Pak, J.H., and Shin, H.J. (2006). *Naegleria fowleri*: functional expression of the Nfa1 protein in transfected *Naegleria gruberi* by promoter modification. Exp. Parasitol. 112, 115-120.

Song, K.J., Song, K.H., Na, B.K., Kim, J.H., Kwon, D., Park, S., Pak, J.H., Im, K.I., and Shin, H.J. (2007). Molecular cloning and characterization of a cytosolic heat shock protein 70 from *Naegleria fowleri*. Parasitol. Res. 100, 1083-1089.

Song, K.J., Song, K.H., Kim, J.H., Sohn, H.J., Lee, Y.J., Park, C.E., and Shin, H.J. (2008). Heat shock protein 70 of *Naegleria fowleri* is important factor for proliferation and *in vitro* cytotoxicity. Parasitol. Res. 103, 313-317.

Song, K.J., Jang, Y.S., Lee, Y.A., Kim, K.A., Lee, S.K., and Shin, M.H. (2011). Reactive oxygen species-dependent necroptosis in Jurkat T cells induced by pathogenic free-living *Naegleria fowleri*. Parasite Immunol. 33, 390-400.

Soto, J., Arana, B.A., Toledo, J., Rizzo, N., Vega, J.C., Diaz, A., Luz, M., Gutierrez, P., Arboleda, M., Berman, J.D., Junge, K., Engel, J., and Sindermann, H. (2004). Miltefosine for new world cutaneous leishmaniasis. Clin. Infect. Dis. 38, 1266-1272.

Sparagano, O. (1993). Differentiation of *Naegleria fowleri* and other *Naegleria*e by polymerase chain reaction and hybridization methods. FEMS Microbiol. Lett. 110, 325-330.

Sparagano, O. (1993). Detection of *Naegleria fowleri* cysts in environmental samples by using a DNA probe. FEMS Microbiol. Lett. 112, 349-351.

Sparagano, O., Drouet, E., Brebant, R., Manet, E., Denoyel, G.A., and Pernin, P. (1993). Use of monoclonal antibodies to distinguish pathogenic *Naegleria fowleri* (cysts, trophozoites, or flagellate forms) from other *Naegleria* species. J. Clin. Microbiol. 31, 2758-2763.

Sparagano, O., Guicher, C., and Peyrol, S. (1997). *Naegleria fowleri*: location of a specific antigen using an immunogold labeling method. Exp. Parasitol. 85, 299-302.

Starnes, G.L., Coincon, M., Sygusch, J., and Sibley, L.D. (2009). Aldolase is essential for energy production and bridging adhesin-actin cytoskeletal interactions during parasite invasion of host cells. Cell Host Microbe. 5, 353-364.

Stevens, A.R., Tyndall, R.L., Coutant, C.C., and Willaert, E. (1977). Isolation of the etiological agent of primary amoebic meningoencephalitis from artifically heated waters. Appl. Environ. Microbiol. 34, 701-705.

Stevens, A.R., Gallup, E.D., and Willaert, E. (1978). Evaluation of membrane-bound black bodies in trophozoites and cysts of *Naegleria* spp. J. Invertebr. Pathol. 31, 63-76.

Stevens, A.R., Shulman, S.T., Lansen, T.A., Cichon, M.J., and Willaert, E. (1981). Primary amoebic meningo-encephalitis: a report of two cases and antibiotic and immunologic studies. J. Infect. Dis. 143, 193-199.

Su, M.Y., Lee, M.S., Shyu, L.Y., Lin, W.C., Hsiao, P.C., Wang, C.P., Ji, D.D., Chen, K.M., and Lai, S.C. (2013). A fatal case of *Naegleria fowleri* meningoencephalitis in Taiwan. Kor. J. Parasitol. 51, 203-206.

Suh, M.R., Han, J.W., No, Y.R., and Lee, J. (2002). Transient concentration of a gamma-tubulin-related protein with a pericentrin-related protein in the formation of basal bodies and flagella during the differentiation of *Naegleria gruberi*. Cell Motil. Cytoskeleton 52, 66-81.

Sussman, D.J., Sellers, J.R., Flicker, P., Lai, E.Y., Cannon, L.E., Szent-Györgyi, A.G., and Fulton, C. (1984). Actin of *Naegleria gruberi*. Absence of N tau-methylhistidine. J. Biol. Chem. 259, 7349-7354.

Sussman, D.J., Lai, E.Y., and Fulton, C. (1984). Rapid disappearance of translatable actin mRNA during cell differentiation in *Naegleria*. J. Biol. Chem. 259, 7355-7360.

Sykora, J.L., Keleti, G., and Martinez, A.J. (1983). Occurrence and pathogenicity of *Naegleria fowleri* in artificially heated waters. Appl. Environ. Microbiol. 45, 974-979.

Symmers, W.C. (1969). Primary amoebic meningoencephalitis in Britain. Br. Med. J. 4(5681), 449-454.

Tang, Y., Nielsen, H., Birgisdottir, A.B., and Johansen, S. (2011). A natural fast-cleaving branching ribozyme from the amoeboflagellate *Naegleria pringsheimi*. RNA Biol. 8, 997-1004.

Taylor, P.W. (1977). Isolation and experimental infection of free-living amoebae in freshwater fishes. J. Parasitol. 63, 232-237.

Tew, J.G., Burmeister, J., Greene, E.J., Pflaumer, S.K., and Goldstein, J. (1977). A radioimmunoassay for human antibody specific for microbial antigens. J. Immunol. Methods 14, 231-241.

Thom, S., Warhurst, D., and Drasar, B.S. (1992). Association of *Vibrio cholerae* with fresh water amoebae. J. Med. Microbiol. 36, 303-306.

Thong, Y.H., Rowan-Kelly, B., Shepherd, C., and Ferrante, A. (1977). Growth inhibition of *Naegleria fowleri* by tetracycline, rifamycin, and miconazole. Lancet 2(8043), 876.

Thong, Y.H., Rowan-Kelly, B., Shepherd, C., and Ferrante, A. (1978). Synergism between tetracycline and amphotericin B in experimental amoebic meningoencephalitis. Med. J. Aust. 1, 663-664.

Thong, Y.H., Shepherd, C., Ferrante, A., and Rowan-Kelly B. (1978). Protective immunity to *Naegleria fowleri* in experimental amoebic meningoencephalitis. Am. J. Trop. Med. Hyg. 27, 238-240.

Thong, Y.H., Rowan-Kelly, B., and Ferrante, A. (1979). Treatment of experimental *Naegleria* meningoencephalitis with a combination of amphotericin B and rifamycin. Scand. J. Infect. Dis. 11, 151-153.

Thong, Y.H., Ferrante, A., Rowan-Kelly, B., and O'Keefe, D. (1980). Immunization with live amoebae, amoebic lysate and culture supernatant in experimental *Naegleria* meningoencephalitis. Trans. R. Soc. Trop. Med. Hyg. 74, 570-576.

Thong, Y.H., Carter, R.F., Ferrante, A., and Rowan-Kelly, B. (1983). Site of expression of immunity to *Naegleria fowleri* in immunized mice. Parasite Immunol. 5, 67-76.

Thong, Y.H., and Ferrante, A. (1986). Migration patterns of pathogenic and nonpathogenic *Naegleria* spp. Infect. Immun. 51, 177-180.

Tiewcharoen, S., Junnu, V., and Chinabut, P. (2002). *In vitro* effect of antifungal drugs on pathogenic *Naegleria* spp. Southeast Asian J. Trop. Med. Public Health 33, 38-41.

Tiewcharoen, S., Junnu, V., Suvoutho, S. (2003). Effect of antifungal drugs on pathogenic *Naegleria* spp isolated from natural water sources. J. Med. Assoc. Thai. 86, 876-882.

Tiewcharoen, S., Komalamisra, N., and Junnu, V. (2004). Zymogram patterns of *Naegleria* spp isolated from natural water sources in Taling Chan district, Bangkok. Southeast Asian J. Trop. Med. Public Health 35, 275-280.

Tiewcharoen, S., Junnu, V., Sassa-deepaeng, T., Waiyawuth, W., Ongrothchanakul, J., and Wankhom, S. (2007). Analysis of the 5.8S rRNA and internal transcribed spacers regions of the variant *Naegleria fowleri* Thai strain. Parasitol. Res. 101, 139-143.

Tiewcharoen, S., Malainual, N., Junnu, V., Chetanachan, P., and Rabablert, J. (2008). Cytopathogenesis of *Naegleria fowleri* Thai strains for cultured human neuroblastoma cells. Parasitol. Res. 102, 997-1000.

Tiewcharoen, S., Rabablert, J., Chetanachan, P., Junnu, V., Worawirounwong, D., and Malainual, N. (2008). Scanning electron microscopic study of human neuroblastoma cells affected with *Naegleria fowleri* Thai strains. Parasitol. Res. 103, 1119-1123.

Tiewcharoen, S., Phurttikul, W., Rabablert, J., Auewarakul, P., Roytrakul, S., Chetanachan, P., Atithep, T., and Junnu, V. (2014). Effect of synthetic antimicrobial peptides on *Naegleria fowleri* trophozoites. Southeast Asian J. Trop. Med. Public Health 45, 537-546.

Tilak, R., Tilak, V., Singh, D.P., Rai, M., Agarwal, S.S., and Gulati, A.K. (2008). Primary amoebic meningoencephalitis relatively undiagnosed public health problem. Indian J. Prev. Soc. Med. 39, 178-180.

Todd, S.R. (1972). Effects of high hydrostatic pressure on transformation in *Naegleria gruberi*. Symp. Soc. Exp. Biol. 26, 485-486.

Toney, D.M., and Marciano-Cabral, F. (1992). Alterations in protein expression and complement resistance of pathogenic *Naegleria* amoebae. Infect. Immun. 60, 2784-2790.

Toney, D.M., and Marciano-Cabral, F. (1994). Membrane vesiculation of *Naegleria fowleri* amoebae as a mechanism for resisting complement damage. J. Immunol. 152, 2952-2959.

Toney, D.M., and Marciano-Cabral, F. (1994). Modulation of complement resistance and virulence of *Naegleria fowleri* amoebae by alterations in growth media. J. Eukaryot. Microbiol. 41, 337-343.

Trimbur, G.M., Goeckeler, J.L., Brodsky, J.L., and Walsh, C.J. (1999). Cloning, sequencing, and nucleolar targeting of the basal-body-binding nucleolar protein BN46/51. J. Cell Sci. 112, 1159-1168.

Tsaousis, A., Nyvltová, E., Sutak, R., Hrdy, I., and Tachezy, J. (2014). A nonmitochondrial hydrogen production in *Naegleria gruberi*. Genome Biol. Evol. 6, 792-799.

Tyndall, R.L., and Domingue, E.L. (1982). Cocultivation of *Legionella pneumophila* and free-living amoebae. Appl. Environ. Microbiol. 44, 954-959.

Tyndall, R.L., Ironside, K.S., Metler, P.L., Tan, E.L., Hazen, T.C., and Fliermans, C.B. (1989). Effect of thermal additions on the density and distribution of thermophilic amoebae and pathogenic *Naegleria fowleri* in a newly created cooling lake. Appl. Environ. Microbiol. 55, 722-732.

van Belkum, A., De Jonckheere, J., and Quint, W.G. (1992). Genotyping *Naegleria* spp. and *Naegleria fowleri* isolates by interrepeat polymerase chain reaction. J. Clin. Microbiol. 30, 2595-2598.

Vargas-Zepeda, J., Gomez-Alcala, A.V., Vasquez-Morales, J.A., Licea-Amaya, L., De Jonckheere, J.F., and Lares-Villa, F. (2005). Successful treatment of *Naegleria fowleri* meningoencephalitis by using intravenous amphotericin B, fluconazole and rifampicin. Arch. Med. Res. 36, 83-86.

Vernhes, M.C., Benichou, A., Pernin, P., Cabanes, P.A., and Teissié, J. (2002). Elimination of free-living amoebae in fresh water with pulsed electric fields. Water Res. 36, 3429-3438.

Viriyavejakul, P., Rochanawutanon, M., and Sirinavin, S. (1997). *Naegleria* meningomyeloencephalitis. Southeast Asian J. Trop. Med. Public Health 28, 237-240.

Visvesvara, G.S., and Callaway, C.S. (1974). Light and electron microsopic observations on the pathogenesis of *Naegleria fowleri* in mouse brain and tissue culture. J. Protozool. 21, 239-250.

Visvesvara, G.S., and Healy, G.R. (1975). Comparative antigenic analysis of pathogenic and free-living *Naegleria* species by the gel diffusion and immunoelectrophoresis techniques. Infect. Immun. 11, 95-108.

Visvesvara, G.S., and Healy, G.R. (1980). Disc electrophoretic patterns of esterase isoenzymes of *Naegleria fowleri* and *N. gruberi*. Trans. R. Soc. Trop. Med. Hyg. 74, 411-412.

Visvesvara, G.S., Peralta, M.J., Brandt, F.H., Wilson, M., Aloisio, C., and Franko, E. (1987). Production of monoclonal antibodies to *Naegleria fowleri*, agent of primary amoebic meningoencephalitis. J. Clin. Microbiol. 25, 1629-1634.

Visvesvara, G.S., and Moura, H. (2006). Pathogenic and opportunistic free-living amoebas: *Acanthamoeba* spp., *Balamuthia mandrillaris*, *Naegleria fowleri*, and *Sappinia diploidea*. Tropical Infectious Diseases: Principles, Pathogens and Practice., ed. R.L. Guerrant, D.H. Walker, and P.F. Weller. Vol. 2. Philadelphia: Churchill Livingstone (Elsevier).

Visvesvara, G.S. (2007). Pathogenic and opportunistic free-living amoebae. Manual of Clinical Microbiology, ed. P.R. Murray and E.J. Baron. Washington, DC: ASM Press.

Visvesvara, G.S. (2010). Amoebic meningoencephalitides and keratitis: challenges in diagnosis and Treatment. Curr. Opin. Infect. Dis. 23, 590-594.

Visvesvara, G.S. (2013). Infections with free-living amoebae. Handb. Clin. Neurol. 114, 153-168.

Vyas, I.K., Jamerson, M., Cabral, G.A., and Marciano-Cabral, F. (2015). Identification of Peptidases in Highly Pathogenic vs. Weakly Pathogenic *Naegleria fowleri* Amoebae. J. Eukaryot. Microbiol. 62, 51-59.

Wade, J., and Satir, P. (1968). The effect of mercaptoethanol on flagellar morphogenesis in the amoeboflagellate *Naegleria gruberi* (Schardinger). Exp. Cell Res. 50, 81-92.

Walochnik, J., Muller, K.D., Aspock, H., and Michel, R. (2005). An endocytobiont harbouring *Naegleria* strain identified as *N. clarki* De Jonckheere, 1994. Acta Protozool. 44, 301-310.

Walsh, C., and Fulton, C. (1973). Transcription during cell differentiation in *Naegleria gruberi*. Preferential synthesis of messenger RNA. Biochim. Biophys. Acta 312, 52-71.

Walsh, C. (1980). Appearance of heat shock proteins during the induction of multiple flagella in *Naegleria gruberi*. J. Biol. Chem. 255, 2629-2632.

Walsh, C. (1984). Synthesis and assembly of the cytoskeleton of *Naegleria gruberi* flagellates. J. Cell Biol. 98, 449-456.

Walsh, C.J. (2007). The role of actin, actomyosin and microtubules in defining cell shape during the differentiation of *Naegleria* amoebae into flagellates. Eur. J. Cell Biol. 86, 85-98.

Walsh, C.J. (2012). The structure of the mitotic spindle and nucleolus during mitosis in the amoebo-flagellate *Naegleria*. PLoS One 7, e34763.

Wang, L., Zhan, Y., Song, E., Yu, Y., Jiu, Y., Du, W., Lu, J., Liu, P., Xu, P., and Xu, T. (2011). HID-1 is a peripheral membrane protein primarily associated with the medial- and trans- Golgi apparatus. Protein Cell 2, 74-85.

Warhurst, D.C., Roome, A.P., and Saygi, G. (1970). *Naegleria* sp. from human cerebrospinal fluid. Trans. R. Soc. Trop. Med. Hyg. 64, 19-20.

Warhurst, D.C., Carman, J.A., and Mann, P.G. (1980). Survival of *Naegleria fowleri* cysts at 4°C for eight months with retention of virulence. Trans. R. Soc. Trop. Med. Hyg. 74, 832.

Weik, R.R., and John, D.T. (1977). Cell size, macromolecular composition, and O2 consumption during agitated cultivation of *Naegleria gruberi*. J. Protozool. 24, 196-200.

Weik, R.R., and John, D.T. (1978). Macromolecular composition and nuclear number during growth of *Naegleria fowleri*. J. Parasitol. 64, 746-747.

Weik, R.R., John, D.T. (1979). Cell and mitochondria respiration of *Naegleria fowleri*. J. Parasitol. 65, 700-708.

Wellings, F.M., Amuso, P.T., Chang, S.L., and Lewis, A.L. (1977). Isolation and identification of pathogenic *Naegleria* from Florida lakes. Appl. Environ. Microbiol. 34, 661-667.

Werth, J.M., and Kahn, A.J. (1967). Isolation and preliminary chemical analysis of the cyst wall of the amoeba-flagellate *Naegleria grüberi*. J. Bacteriol. 94, 1272-1274.

Wessberg, K.L., Skolnick, S., Xu, J., Marciano-Cabral, F., and Kemp, R.G. (1995). Cloning, sequencing and expression of the pyrophosphate-dependent phosphofructo-1-kinase from *Naegleria fowleri*. Biochem. J. 307, 143-149.

Whiteman, L.Y., and Marciano-Cabral, F. (1987). Susceptibility of pathogenic and nonpathogenic *Naegleria* spp. to complement-mediated lysis. Infect. Immun. 55, 2442-2447.

Whiteman, L.Y., and Marciano-Cabral, F. (1989). Resistance of highly pathogenic *Naegleria fowleri* amoebae to complement-mediated lysis. Infect. Immun. 57, 3869-3875.

Wikmark, O.G., Einvik, C., De Jonckheere, J.F., and Johansen, S.D. (2006). Short-term sequence evolution and vertical inheritance of the *Naegleria* twin-ribozyme group I intron. BMC Evol. Biol. 6, 39.

Wildschutte, H., and Lawrence, J.G. (2007). Differential *Salmonella* survival against communities of intestinal amoebae. Microbiol. 153, 1781-1789.

Willmer, E.N. (1956). Factors Which Influence the Acquisition of Flagella by the Amoeba, *Naegleria gruberi*. J. Exp. Biol. 33, 583-603.

Willmer, E.N. (1958). Further observations on the metaplasia of an amoeba, *Naegleria gruberi*. J. Embryol. Exp. Morphol. 6, 187-214.

Woodworth, T.W., Keefe, W.E., and Bradley, S.G. (1982). Characterization of proteins in flagellates and growing amoebae of *Naegleria fowleri*. J. Bacteriol. 150, 1366-1374.

Woodworth, T.W., John, D.T., and Bradley, S.G. (1982). Biological factors affecting enflagellation of *Naegleria fowleri*. J. Bacteriol. 152, 803-808.

Xinyao, L., Miao, S., Yonghong, L., Yin, G., Zhongkai, Z., Donghui, W., Weizhong, W., and Chencai, A. (2006). Feeding characteristics of an amoeba (Lobosea: *Naegleria*) grazing upon cyanobacteria: food selection, ingestion and digestion progress. Microb. Ecol. 51, 315-325.

Yasukawa, H., Sato, A., Kita, A., Kodaira, K., Iseki, M., Takahashi, T., Shibusawa, M., Watanabe, M., and Yagita, K. (2013). Identification of photoactivated adenylyl cyclases in *Naegleria australiensis* and BLUF-containing protein in *Naegleria fowleri*. J. Gen. Appl. Microbiol. 59, 361-369.

Young, M.D., Willaert, E., Neal, F.C., Simpson, C.F., and Stevens, A.R. (1980). Experimental infection of sheep with *Naegleria fowleri* of human origin. Am. J. Trop. Med. Hyg. 29, 476-477.

Young, J.D., and Lowrey, D.M. (1989). Biochemical and functional characterization of a membrane-associated pore-forming protein from the pathogenic amoeboflagellate *Naegleria fowleri*. J. Biol. Chem. 264, 1077-1083.

Yu, Y., Wang, L., Jiu, Y., Zhan, Y., Liu, L., Xia, Z., Song, E., Xu, P., and Xu, T. (2011). HID-1 is a novel player in the regulation of neuropeptide sorting. Biochem. J. 434, 383-390.

Yuyama, S. (1971). The effects of selected chemical agents on the amoeba-flagellate transformation in *Naegleria gruberi*. J. Protozool. 18, 337-343.

Yuyama, S., and Corff, S. (1978). Differentiation-dependent decline of DNA synthetic activities in *Naegleria gruberi*. J. Protozool. 25, 408-415.

Zhang, L., Marciano-Cabral, F., and Bradley, S.G. (1988). Effects of cyclophosphamide and a metabolite, acrolein, on *Naegleria fowleri in vitro* and *in vivo*. Antimicrob. Agents Chemother. 32, 962-965.

Zhou, L., Sriram, R., Visvesvara, G.S., and Xiao, L. (2003). Genetic variations in the internal transcribed spacer and mitochondrial small subunit rRNA gene of *Naegleria* spp. J. Eukaryot. Microbiol. 50, 522-526.

Zysset-Burri, D.C., Müller, N., Beuret, C., Heller, M., Schürch, N., Gottstein, B., and Wittwer, M. (2014). Genome-wide identification of pathogenicity factors of the free-living amoeba *Naegleria fowleri*. BMC Genomics 15, 496.

www.ingramcontent.com/pod-product-compliance
Ingram Content Group UK Ltd.
Pitfield, Milton Keynes, MK11 3LW, UK
UKHW022034041026
15591UKWH00012B/670

9 781910 1905